# IT COULD HAPPEN HERE

# IT COULD HAPPEN HERE

## WHY AMERICA IS TIPPING FROM HATE TO THE UNTHINKABLE— AND HOW WE CAN STOP IT

## JONATHAN GREENBLATT

MARINER BOOKS

*An Imprint of* HarperCollins*Publishers*

*Boston    New York*

marinerbooks.com

Library of Congress Cataloging-in-Publication Data has been applied for.
ISBN 978-0-358-61728-0 (hardcover)
ISBN 978-0-358-62337-3 (e-book)
ISBN 978-0-358-73274-7 (audio)

Book design by Chloe Foster

1 2021
4500842651

*Dedicated to Bernard Greenblatt, my paternal grandfather,*

*whose bravery and courage continue to inspire me every day*

# CONTENTS

## AUTHOR'S NOTE

In writing this book, I've aimed to give you your own ADL handbook against hate, filled with the wisdom and tools you need to identify and push back on prejudice. Accordingly, I've drawn heavily on blog postings, handouts, frameworks, speeches, and other materials from the ADL library. Although I've often indicated in the endnotes where I've adapted language and concepts from previously published material, I've also freely borrowed text from ADL without attribution. I hope you'll find the final product not only readable but also accessible, engaging, informative, and inspiring.

# INTRODUCTION

I'd dreamed of this place for so long. It was charming and quaint in my mind, with cobblestoned streets and green, well-kept parks. Now, gazing around on the train platform, I found a cold and desolate winter landscape—soulless brick industrial buildings, steam rising from nearby chimneys, graffitied walls, just a few bare trees, even fewer pedestrians.

The platform itself was empty save for two gangly soldiers at the far end, guns slung over their shoulders, a light snow falling onto their thick, green Soviet-style overcoats. I made for a small white wooden box of a building about a hundred yards from the station. It wasn't much bigger than an average suburban garage. A sign indicated it was the visitors' center, but that seemed almost ironic. I couldn't imagine many tourists coming here.

Two women sat inside at the counter, one middle-aged, the other elderly. They clutched their sweaters around them against the draft as I closed the door behind me. There were some shelves that held books in German, but otherwise the room was bare.

I addressed the middle-aged woman. "Excuse me, *bitte. Gute tag.* Hello." Using hand gestures and speaking slowly in English, I communicated my request. "My grandfather is from this town. I'm from America. He passed away, but I'm interested in learning about his life,

INTRODUCTION

so I've come here. I'd love to meet people from the Jewish community. I'd love to see the synagogue."

Blank stare.

I tried again, this time pulling out the pocket German-English dictionary I had picked up in Berlin the day before. It was thick with a blue and orange cover. I hadn't even cracked the spine.

"My, uh, *Großvater,* he's Magdeburg, *hier,* here he grew up. Jewish. *Jude.*"

She squinted. *"Dein Großvater? Jude?"*

We went back and forth like this for a while, and she seemed to grow frustrated. The elderly woman asked her a question, and for the next few minutes they conversed with each other in rapid-fire German, pointing and nodding at me. Finally, the older woman turned to me and in heavily accented English said, "Your question, we no understand. Nobody has ever asked it. There are no Jews here. What do you want?"

*There are no Jews here.*

I knew this, of course, but hearing it out loud felt like a punch in the face. I stared back at her, sadness slowly enveloping me.

Many Americans think it unimaginable that hate will ever come for us or our loved ones. We live in the United States, after all, a place of laws and human rights and democracy. Nothing like what happened to the Jews in Germany could ever happen here.

My grandfather used to think that way. He rarely spoke of his youth beyond a passing comment here and there. But a few years before he died, we sat down at my kitchen table and I probed him about his past for a school project. I held a mini–tape recorder in my hand and, trying to understand his life, asked him question after question with the overeager enthusiasm of a high-school junior. Finally, I asked whether, as a young person, he could ever have imagined his grandsons would be Americans, not Germans.

He smiled at me, shook his head, and in his thick accent admitted he never could have. He was born and raised in Magdeburg. He grew up riding his bicycle on its cobblestoned streets. He boxed in the town's youth league and was a fierce player on the soccer pitches. Magdeburg and Germany were all he knew.

"Where else would we be?" he said, sighing.

Hate doesn't devastate its victims right away. You might not feel particularly threatened where you live. But make no mistake—hate can come for you. Half a century from now, *your* grandchild might visit the town where you currently live, inquire about you and your community, and receive nothing but a blank stare.

## One Demagogue Away from Disaster

The organization I lead, the Anti-Defamation League (ADL), is the world's oldest organization dedicated to fighting hate in all its forms. The catalyst for its founding in 1913 was a brutal episode of anti-Jewish hate, the infamous Leo Frank affair. At a pencil factory outside Atlanta, a young girl was raped and murdered. Frank, the factory's Jewish manager, was accused and, despite exculpatory evidence, wrongly convicted of the crime. After his death sentence was commuted to life imprisonment, an enraged mob lynched him. At the turn of the twentieth century, antisemitism was a staple of American life. Jews routinely were maligned in the press and discriminated against in public life. The injustice done to Frank galvanized members of the Jewish community to do something about it.

Remarkably, though, the ADL's founders didn't limit their scope solely to Jews. The organization's original charter calls on the ADL to secure justice and fair treatment for all. Stopping the defamation of Jews was the ADL's "immediate object," but the organization's larger purpose was to "secure justice and fair treatment to all citizens alike and to put an end forever to unjust and unfair discrimination against and ridicule of any sect or body of citizens." The founders of ADL believed in the simple but powerful premise that America could not be safe for its Jews unless it was safe for all its people.

Today my team and I are on the front lines of the global fight against hate, tracking invective and violence and working with law enforcement to prevent tragedies. The trends we're seeing are alarming. Hate is on the rise everywhere, much more than many people realize. Between 2015 and 2018, the United States saw a *doubling* of antisemitic incidents. In 2019, the United States saw more antisemitic incidents than it had in any year in the past *four decades*.

The individuals behind antisemitic incidents subscribe to a range of ideologies. A white supremacist perpetrated the April 2019 attack on a synagogue in Poway, California, while Black Hebrew Israelites shot up a New Jersey kosher supermarket in December 2019. In 2020, QAnon-inspired candidates spouting antisemitic conspiracy theories were elected to the U.S. Congress, and antisemitic imagery was on startling display among perpetrators of the January 6, 2021, attack on the U.S. Capitol.

In recent years, so-called activists hostile to the State of Israel and sympathetic to the Palestinian cause held rallies that unapologetically banned "cops and Zionists." Although banning Zionists might not sound like antisemitism to some, it most certainly is. Since a strong majority of American Jews regard the State of Israel in favorable terms and most American Jews feel a bond with the State of Israel as part of their Jewish identities, banning Zionists from a rally is tantamount to saying *Jews don't belong here*. Moreover, while impassioned criticism of Israeli policies is reasonable, a seething and obsessive hostility toward the world's only Jewish state and its supporters becomes almost indistinguishable from outright hostility toward the Jewish people.

Relatedly, when fighting broke out between Israel and Hamas during the spring of 2021, so-called activists around the world all too often deployed rhetorical violence against the Jewish state and its supporters by, for example, equating Israel and Zionists with Nazis, calling for Israel to be eliminated, and directing anti-Israel messaging at synagogues and other Jewish institutions. That rhetoric in turn likely helped trigger a frightening spike in real-world violence against Jewish people in the United States and around the world.

But hate today isn't just about antisemitism. Look back at history, and you find that Jews are the canary in the coal mine. As Emory University professor Deborah Lipstadt has suggested, hatred starts with the Jews but it often doesn't end with them. The Spanish Inquisition began with attacks on Jews — they were accused of heresy — yet the madness eventually engulfed the entire country. Likewise, the Nazis initially focused on resolving "the Jewish Question" through a campaign of marginalization, persecution, and extermination but soon widened their mandate to encompass other so-called undesirables, including Roma, homosexuals, and the mentally disabled. Eventually, the fire of hatred came to incinerate nearly all of Europe.

The sad fact is that hatred of *all* kinds — including racism, antisemitism, Islamophobia, homophobia, xenophobia, and more — has exploded in recent years. In 2019, the United States saw a reported 7,314 hate crimes — over twenty each day. In 2020, hate crimes against Asian-Americans skyrocketed by almost 150 percent in large urban areas. The problem is especially bad online. A 2021 ADL survey found that 44 percent of Americans had experienced online harassment and that members of marginalized communities reported increased harassment.

And the rise of hate is a truly global phenomenon. In Asia, invective directed at religious minorities — Muslims in India, the Rohingya in Myanmar, Shia in Pakistan — has led to despicable acts of violence. In Africa, displaced peoples have crossed borders seeking refuge from climate-spawned disasters or political unrest only to find themselves subject to racial violence. In the Middle East, long-standing religious hostility has exploded into systemic persecution and even armed conflict, with genocidal consequences for demographic and religious populations like the Yezidis in Syria, the Baha'is in Iran, and other minorities in the region.

In Eastern Europe, nationalist political parties led by demagogic leaders have riled up hatred against refugees and migrants. In Western Europe, the leaders of left-leaning parties like the Labour Party in the United Kingdom and Podemos in Spain have invoked vicious antisemitic conspiracy theories. The Labour Party's Jeremy Corbyn gained global notoriety in the mid-2010s as a politician on the rise who was poised to take 10 Downing Street. His populism was animated by a palpable anti-Jewish hostility. Among many gems Corbyn wrote, he authored the foreword for a book claiming that Jews controlled global finance and trafficking in conspiracy theories. He once said that Zionists, despite "having lived in this country for a very long time, probably all their lives . . . don't understand English irony," a comment widely interpreted as implying that Jewish citizens were not truly British.

Despite Corbyn's protestations that he was not antisemitic, a critical mass of British Jews—almost 40 percent—said in 2018 that they would emigrate if their country elected him prime minister. As they perceived his statements and as they played out in local Labour Party meetings, national events, and op-ed pages across the United Kingdom, he had normalized virulent antisemitism at all levels of the Labour Party.

Why are we seeing so much bigotry? There are a number of reasons. Hate has always been with us and arguably is a latent psychological impulse. Social change and instability—political unrest, mass unemployment, the influx of refugees, pandemics, wars, and the like—can awaken and intensify this phenomenon. When humans feel desperate and uncertain and when dominant institutions and systems fail to deliver solutions, we become more vulnerable to insidious scapegoating and the leaders who peddle these theories. Seeking stability and a way to vent our emotions, we look to blame someone or something for our hardships.

Add in a political dimension—a demagogue who riles up passions and extremists who are eager to benefit from the opportunity—and you have textbook conditions for hatred to spread and even explode into violence.

This invidious dynamic has played out recently in the United States. During the years leading up to the Great Recession, big business violated people's trust, sending jobs overseas and holding down wages. Desperate for better economic opportunities, millions of people turned to Washington, DC, for help, to no avail. Resentment festered, and trust in government plummeted to historic lows. Meanwhile, the increasing diversity of the American population heightened anxiety among some whites, raising the specter of a loss of economic and political power.

After 2008, the Tea Party came onto the scene, and some elements in and around the movement laid out wild conspiracies. They claimed, for example, that Obamacare would institute "death panels," bureaucrats who would arbitrarily decide who received lifesaving care and who didn't, and that Barack Obama was a foreigner, born in Kenya and therefore not qualified to serve as president. These and other lies were seeded on social media and cultivated by willing accomplices in the media, such as Fox News. The hysteria escalated, the demonization intensified, and what once had been fierce partisan opinion morphed into full-blown sectarian rage. By the end of the Obama era, this gospel of hate and misinformation had spread to mainstream politicians and media.

Then along came Donald Trump. Despite proclaiming himself "the least racist person," he had a sordid history of making racist statements, including his contention that the five Black and Hispanic youths wrongly accused in the famous Central Park jogger case deserved the death penalty, that Black people were lazy and made for poor accountants, and that President Obama wasn't a native-born American.

As a neophyte politician, Trump made hate a key element of his pitch to voters; later, it was central to his governing style. From that fateful day in July 2015 when he announced his candidacy as he descended his gilded escalator in Trump Tower, he contrived grim visions of criminals and terrorists storming across our southern border aiming to pillage, rape, and murder. Every hostile tweet, every offensive speech, every coarse comment fanned the flames of hate. And the press breathlessly repeated his hateful messages, allowing intolerance to spread through social media virtually unchecked. Extremists took advantage of the political cover Trump offered them not just to commit acts of violence but to enter the political discourse, further normalizing hate.

For five years, it seemed to many that nothing fundamentally changed — that these instances of bigotry were only words, unpleasant and even immoral but not especially harmful to our democratic system. The attack on the Capitol Building and attempted coup on January 6, 2021, was a wakeup call to the imminent threat posed by Far Right extremists and their hateful ideology, a threat not just to Americans but to proponents of liberal democracy around the world.

Trump's bigotry also catalyzed extremism on the opposite end of the political spectrum. Although nothing like Trump's demagoguery exists on the political left, strident voices have arisen there that espouse a different kind of illiberalism. In particular, a small but steady stream of anti-Zionist critique has veered into blatant antisemitism. In 2015, a student group at Stanford University reportedly asked a Jewish candidate for student government point-blank: "Given your strong Jewish identity, how would you vote on divestment [from Israel]?" This question wasn't posed to any other candidate. It was a litmus test for the Jewish candidate only. In 2017, a lecturer at UC Berkeley retweeted an image of an identifiably Jewish individual with the text "I can now kill, rape, smuggle organs & steal the land of Palestinians." In 2020, it emerged that a professor at University of California, Merced, had been posting vehemently antisemitic messages related to Zionists. In one, he tweeted a photo of the "Zionist brain" that included a "world domination lobe" and a "frontal money lobe." Then, after President Biden won the 2020 election, the professor tweeted: "Surprise, surprise!! The entire system in America is controlled by [the] Zionist. Change of president is just a surface polish, change of veneer. Same trash different pile!"

Entrenched hostility toward Jews among some on the left became painfully clear during the spring of 2021 spike in antisemitism mentioned earlier. Over a two-week period, ADL tracked a 75 percent rise in anti-Jewish incidents, from harassment to vandalism to violence. Research conducted at the end of May 2021 found that a majority of Jews in the United States had personally observed antisemitism, either online or off-, sparked by the conflict in the Middle East. But statistics don't convey the viciousness of what Jews experienced. In Los Angeles, a mob waving pro-Palestinian flags attacked a group of Jewish men as they ate dinner at a restaurant. In New York City, a man shouting antisemitic invective attacked a Jewish man on his way to synagogue, kicking and chasing him for blocks. In Miami, men in an SUV shouted slurs and threatened to rape the female members of a Jewish family.

Such episodes evince the same unadulterated anger and hate that we saw when white nationalists marched through Charlottesville with tiki torches. During the spring of 2021, though, a number of prominent elected officials and other public figures on the left seemed either at a loss for words or unable to offer clear, cogent condemnations, often qualifying their statements with critiques of the State of Israel or comments about anti-Palestinian hate. These issues might merit discussion, but not in response to assailants attacking people in broad daylight simply because they are Jewish. By comparison, when Asian-Americans in the United States suffered a wave of violent, ugly assaults starting in 2020, political leaders didn't condemn the attacks while also arguing that China should change its foreign policy or that Uighur rights should be preserved. Such double standards left American Jews feeling wounded at best, alone at worst.

Our society is becoming more vulnerable by the day to hate on both the left and the right. Beset by a pandemic that has devasted communities, unsettled everyday life, and cost millions of jobs, people are on edge, ever more likely to blame the Other, whether it's Jews, immigrants, Blacks, Asians, Latinx, Muslims, members of the LGBTQ community — you name it. Deepening economic inequality magnifies the tension, as does inadequate health care, excessive levels of personal debt, and stresses caused by once-in-a-century natural disasters that now occur every year. In this environment, with hatred seething around us, the arrival of another demagogue — one smarter and more disciplined than Donald Trump — is all it would take to produce an explosion of violence, mass death, and the destruction of our society and democracy.

But another larger-than-life Donald Trump figure is not necessary for us to suffer this horrible outcome. A "softer," more insidious path to cataclysm also seems possible. As of this writing, figures like freshman GOP representative Marjorie Taylor Greene and Congressman Paul Gosar not only repeat hateful QAnon conspiracies and the big lie about the 2020 election but also trumpet hate through efforts like their short-lived America First Caucus. This quixotic effort could have been labeled the "Ku Klux Klan Caucus" because its call for "common respect for uniquely Anglo-Saxon political traditions" directly invokes the screeds of the most extremist elements in American history.

When news of this effort broke in April 2021, it triggered just a few dispassionate tweets from GOP leadership in the House of Representatives. No rejoinders came from other GOP institutions, ex-president

Trump, or other party luminaries. While Greene and Gosar later attempted to distance themselves from the failed caucus, the damage was already done. Yet another norm of our democracy—the notion that naked racism and antisemitism have no place in the workings of Congress—had been shattered. Perhaps instead of combusting all at once, our nation will fall into violence more gradually as seemingly smaller figures operate on the periphery, pushing the envelope, numbing the public, and shifting norms of acceptable conduct until, at a certain point, what once appeared impossible becomes possible.

Consider this: Nobody would have thought that in America, the government would tear infants from the arms of their immigrant parents and ship them across the country or lock them up in camps or both. And yet, that's exactly what happened. Nobody would have guessed that one day people would see brazen and violent attacks on innocent Jews in places like midtown Manhattan and downtown Los Angeles. And yet, that's also exactly what happened. None of us want to believe that America could end up like Germany in the 1930s. As the American author Sinclair Lewis ironically titled his 1935 novel—published before the full horror of Hitler became apparent—*It Can't Happen Here*. Even today, nobody wants to believe that illiberalism, fascism, and violence could unfold on our shores.

But I wrote this book because we must confront that possibility. What might occur if social instability deepens, hateful attitudes become even more pervasive and entrenched, the traditional institutional protections are worn down even more, and a much shrewder demagogue rises to power? Alternatively, what might occur if a series of opportunistic mini-demagogues on either the right or the left attain power in the next decade, eroding our norms one after the other and making hate increasingly palatable until finally the nation simply lacks the guardrails required to protect it from tragedy?

We can't afford to find out.

## Avoiding Disaster

And hopefully, we won't. We can still confront hatred head-on before it destroys us, attacking it at its roots. The good news is that decent, upstanding Americans far outnumber the haters and insurgents. The even better news is that we have the virtual and figurative tools to fight extremist hate and push it back into the sewer where it belongs.

The ADL has developed a powerful understanding of the process by which societies go off the rails and become genocidal. Our model argues that genocide doesn't come out of nowhere but rather originates in a slow, gradual, and insidious spread of hateful ideas. Hate — which I define as antipathy toward individuals or groups based on their identity characteristics, such as race, religion, ethnicity, gender, and so on — takes relatively innocuous forms at first. Over time, it becomes increasingly entrenched and normalized. More devastating acts of violence occur, up to and including genocide, the systematic effort to eliminate an entire people.

With horrifying attacks on minority groups occurring weekly, our society is growing perilously radicalized and hateful. But as terrifying as that is, this model offers reason for hope. As we at the ADL teach, we can stop the worst, most violent expressions of hate by interrupting the process by which individuals and society become radicalized. We can dial back even deeply entrenched hate if we mobilize a combination of education and public advocacy and courageously call out those who perpetuate intolerance, regardless of their political affiliation or supposed moral position.

The growing peril posed by antisemitism today masks a somewhat more complicated picture. Yes, the extremists have grown more emboldened and dangerous, but until recently the bulk of the population had actually moved in the other direction. During the 1930s and 1940s, antisemitism was rampant in the United States, with about 40 percent of the U.S. population subscribing to hateful beliefs, according to one survey. In 1964, ADL research found, 29 percent held strong antisemitic beliefs, defined as subscribing to "six or more common stereotypes about Jews, out of a total of 11 such stereotypes." As of 2020, only 11 percent of the population was antisemitic by this definition.

This positive change was due to a number of factors, including increased Jewish representation in the media and participation in public life, but a combination of advocacy and education on the part of ADL and other community-based organizations also proved pivotal. Although antisemitism is again on the rise across the political spectrum and hateful acts are exploding, we can push back and help a new generation become more respectful, not just of Jews but of all minority and marginalized groups.

I'm particularly hopeful that we can beat back hate because I've seen how societal beliefs in general can and do change for the better. In 2002, my business-school classmate Peter Thum and I started Ethos Brands, the company that created Ethos Water, the goal of which was to make clean water more accessible to disadvantaged kids around the world. At the time, social entrepreneurship was a novelty, and yet in 2005, Starbucks bought Ethos, and it continues to operate it today. More important, social entrepreneurship has mushroomed to an extent I'd never imagined. Today, there are hundreds of thousands of social ventures around the world.

More broadly, the consensus about the role of capitalism in society has shifted dramatically. In a 1970 *New York Times* article, the economist Milton Friedman articulated what would become dogma in corporate boardrooms: the notion that "the social responsibility of business is to increase its profits." This ethic of profit maximization became so entrenched that when Peter and I tried to raise money for Ethos, prospective investors were more confused than captivated. In one memorable exchange, a venture capitalist suggested I join the Peace Corps.

Today, proponents of Friedman's ideas are in retreat. Prominent business leaders understand that business must benefit multiple stakeholders over the long term, not simply make shareholders rich. Dedicating themselves to higher purposes, the world's largest companies now fund social ventures, operate in more-sustainable ways, and take activist stances on social issues. Norms do shift. Progress does happen. The world does become a better place.

But we do have to work at it. My former boss Barack Obama likes to invoke a teaching of Martin Luther King Jr.: "The arc of the moral universe is long, but it bends toward justice." It sounds nice but it's not quite true. The arc *can* bend toward justice, but much of the time it tacks stubbornly toward the status quo. We must be willing to do the work, reaching up with our own hands and wrenching the arc away from stasis and toward a better future. And when the arc seems to be bending away from justice, we have to dig deep, muster even greater resources, and bend it back.

To date, the ADL has taught hundreds of thousands of people how to respond to bigotry, recognize hate groups, stop online hate and bullying, and educate others about the history of intolerance. Typically, our twenty-five regional offices provide this training to schools, law enforcement agencies, and local groups to help them stop hate from spreading. But with the current uncontrolled spread of hate, our previous efforts are no longer enough. We must arm the general public so that *all of us* can mobilize to eliminate intolerance throughout society.

Think of this book as your own personal ADL handbook against hate, filled with the wisdom and tools you need to push back against prejudice and with memorable personal stories, both mine and others' collected by my organization. I realize that this book will not please everyone. ADL has its share of detractors who reflexively assail our activities, typically filtering our work through a partisan lens. For those on the Far Right, ADL is too liberal and too often makes common cause with progressives. For those on the Far Left, we're too conservative and partner too often with those who resist their ideal of progress.

Rather than attempting to mollify those on the extremes, I've aimed this book at the vast majority of people both in America and abroad who reject dogma and realize that we can't divide the world so easily into black and white. Most of us don't go through life seeking to "win" endless ideological arguments. We're simply trying to earn a respectable living, raise our kids, help our neighbors, and conduct ourselves according to a set of core values. We're frightened by the polarization and vicious rhetoric we encounter all around us, and we seek ways to both understand and combat it.

That's where this book comes in. I hope that after reading it, you'll feel empowered to step up and uphold principles of tolerance, civility, and inclusion in the face of rising bigotry whether you're a young-adult activist, a harried parent, an educator, a business manager, a member of the clergy, or simply a concerned citizen. Part 1 of this book will review the existential threat posed by hate and describe why it's much more dangerous than many people think. In part 2, I'll lay out a program all of us can use to fight back in our personal lives, in our workplaces, through political activism, in our kids' schools, and in our faith communities.

As my friend Sacha Baron Cohen said in his address at the 2019 ADL Never Is Now summit, quoting Voltaire, "Those who can make you believe absurdities can make you commit atrocities." January 6 was bad, but it might have been far worse, and we will certainly see even more atrocities if we don't interrupt intolerance. Hate isn't someone else's problem. It's *your* problem. It's *our* problem. We are not *part* of the solution—we *are* the solution. We can all make a difference, working to change people's minds and soften their hardened opinions. This book shows you how.

I've written this book because I'm scared to death of what the future might hold for the Jewish people and *all* people if present trends persist, a point that I will expand on in chapter 6. I feel compelled to do everything in my power to stop the unthinkable from happening. We must all realize how close we are to widespread violence and the disintegration of democracy and civilized society. My family witnessed firsthand the catastrophes that unfettered hate can inflict on the world. Never again, I say. And I mean it.

*Part I*

# THE PYRAMID OF HATE

1

# HATE GONE MAINSTREAM

W hen I joined ADL in July 2015, a certain anxiety lingered in the back of my mind. I imagined, as a kind of nightmare scenario, that early one morning, my phone would buzz me awake. I would fumble for it in the dark and when I groggily answered it, a staffer on the other end would convey the dreadful news: There had been a horrible massacre at a Jewish site somewhere in Europe — a mass shooting at a synagogue in France, perhaps, or a fiendish act at a community center in Germany, or a pogrom in one of Turkey's remaining Jewish neighborhoods.

This scenario wasn't far-fetched. In recent years, Jewish communities across Europe had been terrorized by acts of mass violence linked to Islamist extremism. In January 2015, a self-proclaimed ISIS supporter murdered four Jews in cold blood during a siege at the Hypercacher kosher supermarket in Paris. That same year a man in Copenhagen, enraged by a cartoonist's portrayal of the prophet Muhammad, attacked a synagogue, killing a security guard whose courage prevented even more carnage. In 2014, a veteran of the Syrian civil war returned to Brussels to shoot four people at the city's Jewish Museum. In 2012, a self-proclaimed al-Qaeda enthusiast attacked the Ozar Hatorah school in Toulouse, France, gunning down a rabbi and his two young sons in the schoolyard and mercilessly shooting an eight-year-old girl in the head. None of these attackers questioned their victims — many of whom were elderly or children — about their views on Zionism or the peace process. They attacked innocent civilians simply for the crime of being Jewish.

I feared it was only a matter of time before another, similar horror would befall Europe. But as it turned out, these fears were misplaced. There would be another attack, but it would strike much closer to home than I had ever imagined.

On the morning of Saturday, October 27, 2018, my wife and I attended services at our synagogue — or shul, as some Jews call it. It was an ordinary Shabbat, no special occasion. Our boys had stayed home, and I had brought my phone with me into the sanctuary only so I could check on them after services. To avoid distractions, I had turned my ringer off, placed my phone in my tallis bag (the bag that carries the fringed prayer shawl, or tallis, a Jew traditionally wears to pray), and nestled the bag in the pew before me. The rabbi had just started his sermon when I heard my phone buzz. I ignored it. It buzzed again. And again. And again. "You better answer it," my wife said.

"It can wait," I said. But the buzzing persisted for several minutes, distracting the people around me. Clearly, something was going on.

I grabbed the bag and walked out of the sanctuary. At the door, I removed my tallis, folded it, inserted it into the bag, and retrieved my phone.

As I looked at the screen, I felt my heart in my throat. Multiple news alerts popped up, informing me that a shooting had occurred at a synagogue in Pittsburgh; there were reports of numerous casualties. The details were hazy, and it was unclear whether the incident was ongoing or had ended. I opened Twitter and steeled myself as I read scattered accounts of the shooting. Almost a dozen people across the country had sent me texts. Some were from Jewish supporters, asking me what I knew. Others were from civil rights leaders, expressing sympathy. All were wrenching.

I called George Selim, ADL's senior vice president of programs, to see what he knew. George was our conduit to law enforcement; I had recruited him to ADL the year before, pulling him from the Department of Homeland Security, where he had worked directly for Secretary John Kelly on countering violent extremism. As he now bluntly confirmed, the shooting in Pittsburgh was indeed a developing situation and I should rush to our Manhattan headquarters so we could get organized, assess the intelligence, and prepare a response.

Over the next few hours, the extent of the horror sank in. We learned that a white supremacist named Robert Bowers had rampaged through the Tree of Life synagogue, murdering eleven people and wounding six in the deadliest antisemitic attack in American history. Bowers had previously posted antisemitic rants on social media, speaking of "filthy Evil Jews" and calling Jews "the children of Satan." Bowers also hated immigrants and was particularly enraged that the Hebrew Immigrant Aid Society (HIAS), an NGO focused on aiding immigrants, actively assisted them. "HIAS likes to bring invaders in that kill our people," he posted to Gab, a social media service, minutes before the attack. "I can't sit by and watch my people get slaughtered. Screw your optics, I'm going in." At the Tree of Life, Bowers shouted, "All Jews must die!" as he shot down elderly worshippers who cowered behind the pews or sought refuge in the hall.

For the next couple of days, I worked feverishly, fielding calls nonstop from television producers, print journalists, elected officials, professional staff, and volunteers, all of whom sought to understand how this atrocity could have happened and the impact it would have on our community. Nearly every person I spoke with shared some kind of personal connection to Squirrel Hill, the idyllic neighborhood where the synagogue was located. Maybe it was a relative who lived there or a former teacher, but everyone seemed to know someone. When they asked me what our society could do to stop future attacks, I paused.

There were no simple explanations, no pat answers. The attack was a devastating reminder of the Jewish people's inherent vulnerability.

Almost a week later, I traveled to Pittsburgh to visit the synagogue and grieve with members of the local Jewish community. I stepped off the plane and spotted travelers idly looking at their phones or running off to their gates, seemingly unaware of what had happened. I walked down to baggage claim and met up with Todd Gutnick, a member of ADL's communications team and a Pittsburgh native who had grown up attending Tree of Life. The two of us went out to his car and drove into Squirrel Hill. As we neared the synagogue, I found myself in a bucolic setting with peaceful, tree-lined streets, tidy storefronts, and well-kept old houses with large front porches. It was the kind of wholesome, neighborly place that television's Mr. Rogers might have lived in. In fact, the real-life Fred Rogers *did* live there for many years, just a few blocks from the synagogue, prior to his death in 2003.

We pulled up to the synagogue and were immediately confronted by grim signs of tragedy. News crews stood on the street corners broadcasting to viewers around the world. Yellow tape kept pedestrians from getting too close to the building. At the front of the synagogue, well-wishers had laid bouquets of flowers in the dozens. Someone had erected as a kind of memorial a series of white Stars of David, each bearing the name of a victim. I took in the scene, watching as people walked up to the building and stood before the markers, silently paying their respects. It was hard to process, an almost out-of-body experience. The Tree of Life resembled the synagogue in Connecticut where I had been a bar mitzvah, where my parents had been married, where my family had celebrated countless festivals. How could such an atrocity have happened *here,* in such a sacred space?

I thought back to a visit I had made in January 2016 to the Hypercacher market in Paris. It was one of my first trips abroad as ADL's CEO, and I found it heart-wrenching to walk in the narrow aisles of the store where Jews had been murdered simply for being Jews. An Orthodox rabbi accompanying me on the visit led me downstairs and showed me the walk-in freezer where an African Muslim employee of the store hid Jewish customers during the attack. To honor the victims, the rabbi helped me wrap my left arm and head with *tefillin*\* and led me in prayer. I was overcome with emotion. In that instant, the weight of my responsibility as ADL's CEO became manifest. It was a moment that I knew would stay with me for the rest of my life.

During that same trip, I visited a private Jewish school in a Parisian suburb and witnessed for myself the horrible, longer-term consequences that unfettered hate can have on communities. Whereas Jewish institutions in the United States traditionally have blended easily into their surrounding neighborhoods, with visitors able to come and go at will, this school was barricaded by fifteen-foot-high walls. Members of the French Foreign Legion wearing camouflage and bearing automatic weapons stood guard in front. When I spoke to the school's principal, I learned that his school was thriving, but for a tragic reason: all Jews in the local community had fled the public schools because of harassment and intimidation. Without exception, the students I spoke with planned to leave France following high school. In their view, the country of their birth had become an inhospitable place to live openly as a Jew.

As I stood in front of Tree of Life, I wondered if that European experience presaged a dark future for America's Jews. Would we too have to fortify our schools, markets, and places of worship? Would we also have to retreat from public institutions and shared spaces because of hate? Would we feel vulnerable, exposed, and marginalized in our own

---

\* *Tefillin* are small wooden boxes containing verses from the Torah. Traditionally, a Jewish male wraps them around his arms and head during morning prayers, but it can be meaningful to use them at any time of day.

land to the point that we would be forced to contemplate exile and a better future elsewhere?

Todd tapped me on the shoulder, pulling me from my thoughts. "Jonathan, I'd like you to meet Scott Schubert, chief of police for the city of Pittsburgh." Schubert and I shook hands, and he immediately conveyed his respect for ADL, sharing that he had participated in an ADL seminar on extremism some years earlier. I asked Schubert what he had encountered on viewing the crime scene shortly after the violence stopped. He stammered, so upset that he couldn't answer. All this veteran officer could do was stare at the ground. As he later told me, officers on duty at the station couldn't believe it at first when they heard reports of shooting at the synagogue. It seemed as unfathomable to them in the moment as it did to me now.

Afterward, I attended memorial services at other nearby synagogues. Community members recalled the humanity of those who had perished, among them two developmentally disabled brothers who had served as greeters at Tree of Life and an elderly couple who had died together in the sanctuary.

The level of sorrow was unbearable. All around me, people were sobbing. I sensed they weren't mourning only the deaths of eleven people, as horrifying as that was. They were mourning the loss of something equally precious and ineffable, a sense of innocence. American Jews used to think that nothing like this could ever happen here. In Israel, maybe, or Europe, but not in America. But the attack in Pittsburgh demonstrated that it could happen here. It seemed clear that even if our community didn't become as besieged as French Jews were, no Jewish site in the United States would ever be quite the same again.

But something else seemed clear as well: If Jews weren't safe, then no marginalized or hated group was safe. It was now conceivable that extremists would not only direct hateful speech at perceived outsiders and dehumanize them but also find them in their most sacred spaces and mow them down in broad daylight.

## Haters from Coast to Coast

The years that followed unfortunately validated these fears. Extremists have perpetrated a series of brutal, homicidal attacks across America, targeting not just Jews but other minorities. In April 2019, on the six-month anniversary of the Tree of Life massacre, a teenager attacked an Orthodox congregation in a city in San Diego County, killing a woman. Many more might have died had the shooter's gun not jammed. A few months later, in August 2019, a white supremacist opened fire at a Walmart in El Paso, Texas, killing twenty-two people and injuring twenty-six in what was the deadliest attack on Latinx people in recent American history.

Then in December, two religious fanatics, self-identified members of the fringe Black Hebrew Israelite community, attacked a kosher market in Jersey City, murdering four people. These extremists harbored antisemitic views, among them the notion that *they* were the true Chosen People, not mainstream Jews. That same month, a deranged man drove to Monsey, New York, during Hanukkah and burst into a home where a group of people were lighting a menorah. He stabbed several of them, including a rabbi who later succumbed to his wounds. The assailant wasn't an avowed white supremacist — he was Black — but he wrote about Nazism in his journal and researched the causes of Hitler's antipathy toward the Jews. As far as anyone knew, he had no history of fighting for Palestinian rights, but prior to the attack, he searched online for Zionist temples.

Other countries have also experienced mass-casualty attacks motivated by hate. In March 2019, a white supremacist killed fifty-one worshippers and injured forty-nine during a shooting at a mosque in Christchurch, New Zealand. Deadly attacks by white supremacists have taken place in Canada, Munich, Sweden, and elsewhere. We've also seen numerous Islamist-inspired terror incidents over the decades, most recently the 2016 shooting at the Pulse nightclub in Orlando that killed forty-nine people.

As visible and horrifying as they are, such atrocities don't begin to capture the full reality of our current scourge of hate. For every "big" hate crime you hear about in the media, countless other demeaning acts take place, from the bullying of Jewish students on school playgrounds to harassment of so-called Zionists on college campuses; from the desecration of houses of worship to militant demonstrations in public places; from hateful comments online to the dissemination of white-supremacist propaganda offline; from the vandalism of property, such as with swastikas, to the hateful hijacking of online meetings (so-called Zoom-bombing).

Gregory Ehrie, the ADL's vice president of law enforcement and analysis, is a U.S. Air Force veteran who spent twenty-two years at the FBI and at one point in his career oversaw all domestic-terrorism investigations in the United States and headed the National Joint Terrorism Task Force. Most recently, he ran all FBI investigations in New Jersey as special agent in charge of the Newark field office. Ehrie recalls that back in 2015, when he oversaw terrorism investigations nationally, there were about eight hundred open cases at any given time. As of 2021, there were approximately two thousand, most of which were related to white supremacist and other Far Right groups.

"That increase is nothing we've ever seen in our nation's history," Ehrie says. "And that's just the FBI. If you extrapolate down to the local and state levels, the numbers of investigations also likely have skyrocketed."

The 7,314 hate crimes logged by the FBI in 2019 took place from coast to coast and in almost every state; in essence, America is a society saturated in hate (for a visual of this, see ADL's hate-crime map for that year at https://www.adl.org/adl-hate-crime-map). The ADL's own data shows that the United States saw over eleven thousand incidents of extremism or antisemitism in 2019–2020, again from coast to coast. To understand the magnitude of the problem, I invite you to visit our website and click on our HEAT map, a continuously updated interactive visualization tool that offers twenty years of data broken down by state.

Hate's disturbing spread isn't evident only in the hate-crime data but in the broader profusion of extremist activity. Across society, hate is at an all-time high, with numerous groups emerging across the ideological spectrum. Hate groups are also more emboldened than they've been in recent memory. "There was a reason the Klan wore masks," Ehrie says, "and that was because they weren't accepted by society." Now white supremacists are flying their colors in the streets, unafraid of social censure.

They're also organizing more openly. Military officers and current and former law enforcement officers are not only imbibing hateful ideology but organizing on its behalf and weaponizing it as never before. They are members of traditional groups and of emerging networks like the Proud Boys and QAnon. Although racism and antisemitism have always formed the core of white-nationalist ideology, today it also links up seamlessly with virulent anti-immigrant and anti-Muslim views.

It's hard to underestimate the organizational boost that white supremacists, anti-immigrant groups, neo-Nazis, white nationalists, and others received during the Trump years, when their ideology was tolerated and sometimes openly encouraged by officials at the highest levels. There are almost too many examples to count. Before his election, candidate Trump adopted the America First slogan pioneered by

Charles Lindbergh in the 1930s as a tactic for isolating and marginalizing American Jews. Once Trump came to power, members of his White House staff gave speeches at white-supremacist events. Interns flashed white-power signs in official White House photos. Staffers issued press credentials to Holocaust-denying media outlets. And the president himself made the far-fetched claim that he didn't know anything about QAnon except that its supporters liked him. There is nothing normal about such behavior.

At the same time, we've seen an alarming rise of hateful forces at the extreme end of the political left, forces that often seem to cohere around an irrational and obsessive intolerance toward Israel and that often target all supporters of the Jewish state. I've offered some examples, but there are many more. Just in the past couple of years, a Jewish student leader was compelled to resign from student government following harassment for identifying as Zionist. Musician Roger Waters routinely employs tropes bordering on antisemitic (if not outright antisemitic) in his obsessive focus on Israel. In June 2020, Waters called the late Jewish Republican donor Sheldon Adelson the "puppet master" of U.S. foreign policy and spread the lie that the Israel Defense Forces taught U.S. police the lethal tactics used against George Floyd, a complete fabrication but one regularly employed by some on the Far Left. And radical anti-Israel activist Miko Peled continues to make the rounds on college campuses and in leftist spaces. He has openly trafficked in the antisemitic trope that American Jews are lesser citizens because they harbor dual loyalties to Israel and America. That Peled is an Israeli Jew demonstrates an unfortunate reality: Jews can also legitimize antisemitic tropes.

Even though some earnestly seek to distinguish between anti-Zionism and antisemitism, we must acknowledge the multitude of instances in which anti-Zionism has veered into hatred or discrimination against Jews. To denounce this phenomenon head-on doesn't weaken the Palestinian cause, as some activists claim. On the contrary, those who argue to this effect help lay the groundwork for more anti-Jewish hate.

The truth is that anti-Zionism and antisemitism have long gone hand in hand. In the Soviet Union during the Cold War, as authorities terrorized Jews living within their borders, they pioneered anti-Zionism as a political weapon and deployed it widely in their propaganda, drawing on classic antisemitic tropes in the course of lambasting Israel and its supporters. Russia had a long history of persecuting Jews, but anti-Zionism was a new tool that the Soviet authorities could use to hide their animus. Essentially they were saying, *It's not the Jewish people we hate, it's the Jewish state*. But the end result was the same.

A similar scenario unfolded in much of the Arab world. Jews had lived in these countries since antiquity, undoubtedly as second-class citizens but still in coexistence with the Muslim majority. Ahead of 1948 and after the founding of the State of Israel, Arabs suddenly derided Jews as Zionists and targeted them with a wave of hate. From Iraq to Egypt, frenzied mobs targeted their Jewish neighbors, hounding them, driving them from their homes, destroying their businesses, and slaughtering them in hate-filled pogroms that evoke heinous historical memories familiar to Americans, such as the Tulsa race massacre of 1921 and the Chinese massacre of 1871 in Los Angeles.

Today, antisemitism in all its manifestations is gaining strength. Absent an outcry, radicals on both sides of the ideological spectrum perceive that there's nothing wrong with what they're doing, and they feel like they're being accepted. They sense that they have strength in numbers as long as they stay together.

Looking again at hate in general, we find that many groups are drawing inspiration from one another; they are also sharing knowledge about "best practices." When one group commits a high-profile crime, others seek to understand how they succeeded and try to emulate their methods. The terrorist who attacked a mosque at Christchurch, New

Zealand, livestreamed his atrocities on Facebook, echoing the Islamic State's earlier technique of disseminating videos of murders. A perpetrator who shot up a school in New Mexico was in direct contact with another violent extremist who killed nine people in a Munich, Germany, shopping mall. As such examples suggests, hate groups are forging global connections, even with their ideological antagonists. It might seem puzzling that white supremacists would openly admire Islamic extremists from the Middle East and adopt their tactics, but that's precisely what they're doing, purporting to wage what they term "White Jihad."

In one example, the Base, a violent white-supremacist group active in the United States and elsewhere, borrowed its name and structure from al-Qaeda. The small international neo-Nazi movement known as the Feuerkrieg Division, or FKD, encouraged supporters to make their own explosives by borrowing imagery from a video disseminated by the Islamic State and adding the tagline "It's easier than you think." Another group, the Atomwaffen Division, openly applauded Islamic extremists, arguing that "the culture of martyrdom and insurgency within groups like the Taliban and ISIS is something to admire and reproduce in the neo-Nazi terror movement."

In September 2020, law enforcement caught two adherents of the anti-government movement Boogaloo Bois allegedly attempting to sell weapons to the Palestinian extremist group Hamas. As a U.S. government official explained, "Thinking that they shared the same desire to harm the United States, they sought to join forces and provide support, including in the form of weapons [and] accessories." The extreme right and radical Islamists frequently converge around other key ideological points, such as their hatred of Jews, their misogyny, and the revulsion they feel toward LGBTQ people.

Countries around the world are beginning to designate extremist groups as terrorist threats even if they're not physically based in their territory. Hamas has long been classified as a terror organization by the United States and governments around the world. In 2021, Canada designated the U.S.-based Proud Boys a terrorist group and a threat to its domestic security, adding the Proud Boys to a list that included the white-supremacist groups Atomwaffen Division and the Base and the Russian nationalist group Russian Imperial Movement. It was a striking decision but a logical one, considering the violent track records of these organizations.

## The Disneyfication of Hate

Violent extremist groups' activity is one thing, but we're also besieged by a broader wave of hateful speech and ideology that is infiltrating American culture and daily life. Since 2006, Oren Segal has led the ADL's Center on Extremism; he is charged with tracking all forms of extremism and hate online and in the real world. Segal and his team are experts in the toxic ideologies, obscure symbols, and methods of diverse extremist movements. Segal notes that a profound cultural shift has taken place over the course of a generation. Whereas once, conspiracy theories and extremist movements marked by antisemitism, racism, and other hateful ideas were a somewhat obscure issue, now they're easy to access and readily available, making it more likely that people will fall under their sway.

A toxic combination of social media and demagoguery is to blame. Segal speaks of "a social media phenomenon where we all create our worldview through the information we consume and where conspiracy theories and disinformation are as ubiquitous as legitimate news." The result is a subtle but pervasive *normalization* of hate, made worse by political leaders who disseminate ideas that were formerly beyond the pale, affirming their validity, or who remain silent, giving others an excuse not to step up and correct hateful rhetoric.

Take the classic antisemitic trope of the scheming, greedy Jewish person wielding secret power behind the scenes. Right-wing extremists long portrayed Jewish bankers, such as the Rothschild family or George Soros, as a sinister force, manipulating events to harm the white race. In recent years a growing number of conservatives have adopted this rhetoric, accusing Soros of outlandish crimes like funding a caravan of immigrants to invade the United States and change its culture. On the left, meanwhile, some describe nefarious global schemes by the State of Israel and Zionists to intervene in the political affairs of different countries. Their alleged goal: to foment chaos from which the Jewish state might benefit at the rest of humanity's expense. A couple of decades ago, the vast majority of Americans would never have encountered such ideas. If they had, they would have regarded them as outlandish and obviously false.

Today, these toxins might originate on the margins but they make their way into the mainstream, bubbling up from the recesses of a wide range of unfiltered and unregulated online forums and spaces to private message boards and then onto public social media feeds, where they inexplicably appear alongside postings from legitimate news sites. On the right, for instance, pundits on outlets like Fox News echo hateful ideas, spreading these fictions in their nightly harangues. Political leaders complete the process, normalizing the abnormal with a wink and a nod. President Trump did precisely this in October 2018 when he insinuated to a reporter that "a lot of people" believed Soros was funding migrant caravans and that he "wouldn't be surprised" if these rumors were true. A similar dynamic often plays out on the left. Extremist voices in places like Tehran spew wild accusations of genocide against Israel. Social media accounts launder these accusations, and they subsequently show up on news sites like the *Intercept,* which publishes some serious journalism but also often circulates anti-Israel invective. And then, perhaps unsurprisingly, players in the mainstream political world further legitimize these accusations. When these theories are legitimized through credible public voices, they become weaponized.

Extremist agitators exploit the opening provided by social media, manipulating messaging to change minds and win new adherents. On the one hand, they make hate more attractive to younger audiences by clothing it in ironic humor and embedding it in youth culture. Hatred today takes on the familiar tonality of adolescent transgression, whether communicated by media figures like Alex Jones and Tucker Carlson or by actual teens themselves. Segal notes that his team routinely fields reports of teens posting images on social media of their Jewish classmates in ovens and students bullying immigrant children by confronting them at school and saying, "We're going to build a wall." One vicious hate group mentioned earlier, Feuerkrieg Division, was actually run by a thirteen-year-old from Estonia. The group recruited members online and sought to foment an "apocalyptic race war."

At other times, hate groups soften their ideas so that they might slip more easily into polite discourse. A small but notable segment of commentators on the left (or allied with those on the left) soften their calls to destroy the Jewish state by appropriating the language of civil rights and advocating for justice for the Palestinian people. Iranian influence networks often use this tactic. As an example, one of them, the Iranian government–owned broadcaster Press TV, gave inordinate coverage to the Black Lives Matter protests in the summer of 2020, frequently trying to draw comparisons, no matter how forced, to the situation in the West Bank. Dead wrong on the facts, Press TV applied the Black Lives Matter lens to the complex Middle East situation in a reductive way, treating the Palestinians as the victims of aggression not of racist police but of a supposedly racist Jewish state. There was no acknowledgment of the complexity of the history or the legitimate concerns of both sides; Press TV delivered a contrived conclusion: Israel is an evil oppressor that must be destroyed.

In the United States, anti-Israel and anti-Zionist groups such as American Muslims for Palestine (AMP), Students for Justice in Palestine (SJP), Jewish Voice for Peace (JVP), and If Not Now (INN) (which claims not to have a "unified" stance on Zionism) normalize bigotry in an especially insidious way, spewing toxic rhetoric while engaging in a wide range of otherwise innocuous activities. These groups' ability to harness narratives separate from their most extreme positions allows them to gain and maintain a foothold among more-mainstream progressives. Such organizations appear to some progressives as laudable standard-bearers in combating perceived injustices against Palestinians. Unfortunately, that position and the large networks of grassroots activists maintained by these groups guarantee that when their leaders express inflammatory and sometimes hateful rhetoric, it becomes disseminated widely, poisoning the environment for Jewish communities.

Dr. Osama Abuirshaid, AMP's executive director, has demonized Zionists and Zionism, portraying the latter as "a forged ideology, an ideology that hijacked a religion, an ideology that politicized the religion, an ideology that injected racism into a religion." On one occasion in 2016, he accused Zionists of being unpatriotic Americans because of their "double loyalty" to Israel and America, a classic antisemitic trope. It is hard to imagine progressives tolerating such language directed at other ethnic minorities such as AAPIs, Latinx, or even Arab-Americans, groups who might also harbor deep connections to their homelands.

AMP chairman Hatem Bazian, who cofounded the SJP and serves as a lecturer at University of California, Berkeley, has effectively called for a boycott of much of the Jewish community, repudiating engagement with the ADL and other mainstream Jewish groups as "a trojan horse to normalize Zionism and Israel while building relations with Zionist organizations." In July 2017, he retweeted (and ultimately apologized for) an antisemitic meme depicting an Orthodox Jewish man raising his hands in the air and a caption reading: "Mom, look! I is chosen! I can now kill, rape, smuggle organs & steal the land of Palestinians yay #Ashke-Nazi." In June 2021, SJP targeted ADL's Boston office, humiliating and assaulting an individual who simply observed their deranged extremism. The victim's personal account of the traumatic experience sounds familiar to anyone who has been attacked by a seething mob.

Again, such rhetoric and actions—and this is just a sampling—represents only a fraction of these groups' activities, which include film screenings, so-called cultural heritage events, and (in the case of some Jewish anti-Zionist groups) Shabbat dinners and other religious celebrations. When mainstream Jewish activists object to clearly antisemitic rhetoric, progressives often regard these protestations as efforts to smear Muslims, or they minimize the rhetoric as aberrations. But all expressions of hate matter. Anti-Zionist and anti-Israel groups might not be shooting up synagogues or attacking Jews on the streets, but make no mistake—words can and do beget violence. These groups contribute to a toxic environment in which Jews become marginalized, disrespected, and dehumanized. Gradually and almost imperceptibly, antisemitism may become normalized on the left, creating the potential for such hatred to mutate into even more pernicious and deadlier forms.

We see a similar softening in the rhetoric of right-wing extremism. If you talk openly about lynching Blacks and shooting Jews, you might turn off large segments of the population (thankfully). But your message will resonate with more people if you tone down the language, talk about justice, and even appropriate civil rights vocabulary. One extremist group, the National Socialist Movement, temporarily changed the icon on their flag from a swastika to a Nordic rune. Rather than feeling instantly repelled upon spotting the group's flag, a member of mainstream society might wonder about the symbol's meaning, look it up, and in that way become exposed to hateful ideas. The group actually told its members: "Your Party Platform remains the same, your Party remains unchanged, it is a cosmetic overhaul only."

Although menacing images of heavily armed men in combat garb sometimes define extremism in the public eye, we've also become accustomed to seeing figures like Richard Spencer wearing a suit and tie and yet spewing classic racist and antisemitic ideas, and supposedly respectable academics like the late Tony Martin blaming the Jews for the slave trade and spreading other hateful ideas. Judging from their names, some newer right-wing groups, such as Identity Evropa and the Traditionalist Worker Party, don't seem like white-supremacist organizations at all. Sadly, they most certainly are.

Beyond this Disneyfication of extremism, hate spreads so easily for the same reason it always has — the ideas themselves are at once simple, adaptable, and compelling to certain groups of people. Although the specifics of hateful ideology change over time, the core ideas — a sense of grievance and dehumanization of an evil Other — are nothing new. Strip everything else away, and you wind up with narratives in which the supposed victims are fighting back against seemingly dire, existential threats. White supremacists fault Jews and Blacks for taking away their culture and what they claim as their rightful power. Anti-government activists think liberals are plotting to take away their guns. QAnon members think that a nefarious cabal is stealing their freedom. Some extreme left-wing groups believe that the Jewish state is responsible for all the world's ills. And on and on it goes.

If one specific narrative or argument no longer resonates or becomes dated, another one that seems more relevant pops up. And if you think the obviously nonsensical nature of many hateful ideas will prevent their dissemination, think again. Even patently absurd ideas resonate with certain individuals who are desperately looking for meaning, a sense of belonging, and a way to channel their alienation and anger.

Many of us were shocked by the January 6, 2021, attacks on the Capitol Building. For Segal, it represented the physical embodiment of the mainstreaming of hate he'd been tracking for years. Although some of the rioters were hard-core members of extremist groups, the vast majority of participants were everyday people who didn't belong to hate groups but who had accepted certain of their ideas and been drawn into radical behavior. "I think all of us are susceptible to these narratives," Segal says, "which is why our leaders must constantly remind us that they're either hateful or conspiratorial or false. When we've stopped doing that, then the baseline can animate more people to anger and hatred. I think that's where we're at. The baseline of what is acceptable in terms of hatred has changed."

## Teetering on the Edge

Ehrie, Segal, and others at ADL are frontline warriors in the struggle against hate. Like me, they spend their days confronting humanity's most disturbing and repulsive sides. It can be exhausting, dispiriting, and, frankly, quite scary work. In a later chapter, I'll point to what America might look like if present trends go unchecked. It's a terrifying picture: more violence, political instability, a degraded public discourse, the death of democracy—in short, a world reminiscent of 1930s Europe in which full-blown genocide becomes a real possibility.

At the same time, both Ehrie and Segal feel hopeful that Americans *will* make the right decisions and push back against hate. "I always remind myself of this," Segal says, "that for every shooting or bombing or insurrection or instance of online harassment, there are many stories of people rejecting it and pushing back against it."

Segal is right. As devastating as hate can be, it's not a one-sided story. The more mainstream hate becomes, the more we find good-intentioned people stepping up to counter it, learning from their mistakes, and helping to spread messages of love and peace.

The Tree of Life atrocity, as horrible as it was, offers cause for hope. The wounds left by the assailant may never heal. They remain a source of searing pain for the families of those killed and wounded as well as for the broader community. In the wake of the attack, white supremacists celebrated, spreading the hashtag #HeroRobertBowers on social media and pronouncing it as "another holiday to add to the calendar." Extremists inspired by the attack hatched plots to attack other Jewish targets. Many were thwarted by authorities, but as I've mentioned, one succeeded: the attack on the Chabad synagogue in Poway, California.

At the same time, the Tree of Life tragedy sparked a spontaneous outpouring of sympathy and love. The night after the attack, a vigil for the victims held in Squirrel Hill drew thousands of people. In the days that followed, Pittsburgh's Muslim community showed their support by raising money to pay for funerals of the victims. The Pittsburgh Steelers held a moment of silence for the victims during their game. Businesses provided free goods and services to traumatized residents. An entire community came together in pain and outrage to push back against hate.

Pittsburgh has "changed and . . . will never be the same," Mayor Bill Peduto said in a video commemorating the first anniversary of the attack. "And at the same time, Pittsburgh showed off its best character and in doing so proved itself to be resilient." Dr. Jeffrey Cohen, president of the hospital that treated Robert Bowers after he was wounded by police, agreed. "There's a core decency that exists here that's very hard to shatter. And Mr. Bowers did not do that."

We must all tap into our core decency and stand up against hate whenever and wherever we encounter it. It's easy to respond when we perceive that haters are targeting members of our own tribes, but we must also get in the habit of speaking out and taking action when *anyone* is the target. The presence of hatred anywhere affects us all. It coarsens our discourse. It weakens our democratic institutions. It creates an environment of mistrust that renders every one of us less safe.

When I visited Pittsburgh in the wake of the shooting and stood there among the mourners, I felt hollow and helpless. My job, first and foremost, is to protect the Jewish people. By an objective measure, I'd failed. But I'm not helpless. None of us are. Our society might be teetering on the edge, but there's still time to stand up for the peaceful, democratic society we want to gift to our children and grandchildren. We can secure a future in which we celebrate our common humanity and treat everyone with respect, irrespective of differences.

Let's not wait until a madman spewing extremist ideology shoots up another synagogue, mosque, childcare center, or school. Let's push back against smaller instances of hateful speech and ideas before they have a chance to radicalize more people. The bigoted behavior we encounter in our daily lives or perhaps even engage in ourselves on occasion might seem innocuous, but it's not. In addition to the victims whom it hurts today, such behavior sets the stage for more hate — and more devastating acts of violence — tomorrow. An understanding of this basic reality is one of the most fundamental tools we have at our disposal to counter the hate in our midst.

# 2

# FROM MICROAGGRESSIONS TO GENOCIDE

In March 2021, a video surfaced of Meyers Leonard, a basketball player for the NBA's Miami Heat, using the word *kike*, a deeply derogatory term for Jews, in the course of playing the video game Call of Duty and livestreaming his performance on the online platform Twitch. Leonard was insulting a competitor, as video-game players often do. He blurted out the word in the midst of a string of other profanities.

Thankfully, people noticed. An uproar ensued. Twitch temporarily banned Leonard from its platform, and several gaming companies cut their sponsorship ties with him. The Heat condemned Leonard's comment, and the NBA suspended him for a week and levied a $50,000 fine. Weeks later, the Heat traded him away to the Oklahoma City Thunder.

The day after the incident, Leonard posted an apology on social media, claiming he hadn't known what *kike* meant or that it was offensive and that he was "committed to properly seeking out people who can help educate me about this type of hate and how we can fight it." It turned out I was one of those people.

Shortly after the controversy broke and after the ADL condemned his slur, representatives of Leonard's contacted me to ask if I would take his call. Apparently, he wanted to apologize for his actions. I obliged and we spoke for nearly an hour on a Zoom call, where he was joined by several people.

During our conversation, I felt that Meyers was genuine in his contrition. It seemed clear that he really hadn't intended to cause offense and had uttered his ugly slur from a place of ignorance and without real malice. Nonetheless, he acknowledged that the word still stung for many and that he would need to make amends. We discussed how he could do so, exploring the idea of him working with us to counter hate in the world of online games, where it is rampant and growing. ADL researchers found that almost three-quarters of American adults playing online games were harassed in some way. Over half of those harassed were targeted because of some facet of their identity (race, religion, gender, and so on).

Given the frequency of smaller transgressions like racially insensitive comments and ill-considered ethnic jokes, you might wonder if it's worth calling them out. Aren't we "overreacting" or being "overly sensitive"? Do we really need to embarrass celebrities and athletes when they cross an imaginary line? Is it necessary to condemn every act of hateful bullying on the playground? Should we honestly express outrage when a swastika turns up on a street sign or when someone leaves hate-filled pamphlets on a college campus?

I believe we absolutely must.

If we want to prevent the unthinkable from happening, we can't just fight back against hate crimes and counter organized extremist groups. We also must call out smaller acts of hate whenever we encounter them. That doesn't mean that all teasing is off-limits. But it does mean we should develop some basic judgment about what's inbounds and what's out-of-bounds. Certainly, an influential celebrity using such a harsh, manifestly offensive word or ethnic slur in such a public setting should be out-of-bounds.

Another professional athlete, New England Patriots star wide receiver Julian Edelman, helps us understand why. When Leonard's slur became public, Edelman stepped up to counter hate; he posted an open letter to Leonard on social media that was at once empathetic, forceful, and wise:

> An open letter to Meyers Leonard
>
> So we've never met, I hope we can one day soon. I'm sure you've been getting lots of criticism for what you said. Not trying to add to that, I just want to offer some perspective.
>
> I get the sense that you didn't use that word out of hate, more out of ignorance. Most likely, you weren't trying to hurt anyone or even profile Jews in your comment. That's what makes it so destructive. When someone intends to be hateful, it's usually met with great resistance. Casual ignorance is harder to combat and has greater reach, especially when you command great influence. Hate is like a virus. Even accidentally, it can rapidly spread.
>
> I'm down in Miami fairly often. Let's do a Shabbat dinner with some friends. I'll show you a fun time.
>
> JE

In this brief message, Edelman, who is Jewish, captured something essential about today's mainstreaming of hate. No, Leonard didn't shoot up a synagogue or assault an elderly Asian-American person on the street. He didn't violently harangue members of a marginalized community because of their supposed political views. He blurted out a hateful slur—something that happens countless times a day in the United States and around the world.

But as Edelman correctly points out, it's precisely the casual, everyday nature of the behavior that makes it so damaging. Blatant acts of contempt spark reactions that can serve to somewhat neutralize the damage. Smaller, less obvious acts of hate speech—a put-down, a stupid joke, and the like—slip through unnoticed. Through seemingly inconsequential acts of hate, our discourse becomes imperceptibly degraded and our prejudices become part of everyday reality. Ignorance calcifies as if it were the natural order of things. Hateful beliefs spread, increasing the potential for more harmful speech and even outright violence going forward.

*Every* hateful thought or deed matters—not just because of the direct harm it causes to others, but because of the *indirect* role it plays as a vector for the spread of hate. Horrific attacks such as the one that took place at the Tree of Life synagogue don't just happen out of nowhere. They happen because over time, biased thinking and disrespectful behavior become normalized, leading larger numbers of people to demonize, dehumanize, and diminish the outsider.

Almost inevitably, some people will feel moved to lash out and take action on their prejudices. As small acts of violence become more common, they reinforce the hateful ideas and assumptions they embody and give rise to even bigger, more devastating acts. Gideon Taylor, president of the Claims Conference, reminds us that "the Holocaust started with words. These were hateful words that were yelled in the park, spat on the street, and roared in the classroom. These words alienated, belittled, and shocked; but worse, these words gave birth to the horrific massacre of six million Jews."

## The Pyramid of Hate

At the ADL, we have a powerful model that we've used for decades in our educational programs to help people understand in broad terms the process by which the cancer of bias and hate metastasizes and becomes more dangerous in societies. This model also articulates the relationship between small, seemingly innocuous acts of hate and acts of mass violence. We call it the Pyramid of Hate:

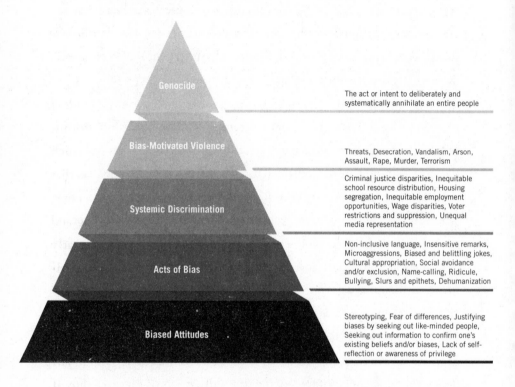

Genocide — The act or intent to deliberately and systematically annihilate an entire people

Bias-Motivated Violence — Threats, Desecration, Vandalism, Arson, Assault, Rape, Murder, Terrorism

Systemic Discrimination — Criminal justice disparities, Inequitable school resource distribution, Housing segregation, Inequitable employment opportunities, Wage disparities, Voter restrictions and suppression, Unequal media representation

Acts of Bias — Non-inclusive language, Insensitive remarks, Microaggressions, Biased and belittling jokes, Cultural appropriation, Social avoidance and/or exclusion, Name-calling, Ridicule, Bullying, Slurs and epithets, Dehumanization

Biased Attitudes — Stereotyping, Fear of differences, Justifying biases by seeking out like-minded people, Seeking out information to confirm one's existing beliefs and/or biases, Lack of self-reflection or awareness of privilege

Let's review each level of this pyramid in turn. *Biased attitudes* are at the bottom of the pyramid and are almost universal. All of us harbor biases, such as assuming that "all Jews" or "all Black people" or "all women" or "all gay people" are a certain way. Biases and stereotyping are natural human tendencies, part of the way we process information and work to simplify a complex world. The cultural milieus in which we grow up—our families, the media, our peers and teachers—also program these beliefs and assumptions into our brains.

As we move through the world, we seek out evidence and like-minded people who validate these biases, and we ignore evidence and people who challenge them—what psychologists call "confirmation bias." In this way, our biases become locked in. We take for granted that they're true and resist it when others point out their falsity.

We also develop implicit biases—quick judgments we unknowingly make about others when we encounter them and that govern our actions. It takes effort to reflect on our beliefs, become aware of our implicit biases, and inject complexity and humanity back into our assessments of others.

As prejudices linger and strengthen, they might give rise to *acts of bias*—the second level of the pyramid. Acts of bias take many forms, including insensitive remarks, ethnic jokes, name-calling, and slurs such as the one used by Meyers Leonard. A glance at the pyramid shows that this category varies widely in terms of the seriousness of the acts, ranging from microaggressions and noninclusive language to bullying and dehumanization, all of which can cause harm. Context matters greatly here, as it influences how targets and victims understand and experience any given act.

Individual acts of bias are bad enough, but hate can become so pervasive in a society that it gives rise to *systemic discrimination*. In the United States and around the world, discrimination becomes enshrined in law and common practice, affecting virtually all aspects of everyday life. If you're Black, for instance, you stand a far greater chance of being arrested and convicted than a non-Black peer and also of being sentenced more harshly. You stand a much greater chance of attending lower-resourced schools, receiving poor health care and dying of underlying conditions, experiencing high levels of environmental pollution, and so on. You have a harder time getting a bank loan and voting, especially as lawmakers make it even more restrictive through laws that by all measures are designed to disenfranchise Black voters.

If you're a woman, you historically earn less than a man for the same work and have a much harder time getting promotions and attaining positions of power in most fields. If you're LGBTQ, you couldn't legally get married and experience the associated rights and privileges of marriage until just a few years ago. In most states, you can still be fired from your job or denied housing based on your sexual orientation or gender identity.

When hate becomes so prevalent that it pervades our very social institutions, the next level of the pyramid — *bias-motivated violence* — isn't much of a stretch. This category also spans a wide array of behavior, from vandalism of property and severe harassment all the way up to murder and acts of mass violence.

At the top of the pyramid, we find the rarest but most catastrophic acts of hate, those conveyed by the term *genocide*. The ADL defines *genocide* as "the act or intent to deliberately and systematically annihilate an entire people." By this standard, the Nazi Holocaust against the Jews during World War II clearly qualifies, but so do a number of other episodes, including the murder of about a million Tutsis in Rwanda, the destruction of minority populations in the former Soviet Union under Joseph Stalin, and the "ethnic cleansing" of Muslims in Bosnia during the 1990s. More recently, the systematic slaughter of the Yezidi minority by ISIS in Syria qualifies as another such act.

## A System of Oppression

As the pyramid suggests, hateful acts of violence don't occur in a vacuum. These acts might be rare compared with levels of biased attitudes and nonviolent hateful acts, but they take place in a larger social context in which biases and nonviolent acts are commonplace. In the absence of biased attitudes and acts, systemic discrimination probably wouldn't be as pervasive, nor would you see as many horrific mass shootings targeting people of color and other marginalized people or as much organized violence against entire groups. This isn't to say that a casual slur "causes" genocide or that the presence of hateful ideas in a society means it will inevitably slide into violence. Rather, we can simply observe that the prevalence of hateful beliefs and actions can spread and escalate over time, becoming normalized and creating conditions that make extreme acts of violence more likely.

No model of reality is perfect, of course, and beyond a point, it can be tricky to pin down relationships of causality. Biased attitudes (level 1) might lead to systemic discrimination (level 3), but discrimination might also spawn biased ideas. For instance, historical biases have kept many women from pursuing careers in the sciences. But some people point to the absence of women in these fields as *evidence* that women can't compare intellectually to men. It's obviously circular logic and flawed, but it still has impact.

We should acknowledge that some concepts cross over the various levels of the pyramid. We might dehumanize others by making disparaging remarks—for example, calling a member of an ethnic group an "animal" or an "alien." But we might also dehumanize others via our actions or systemic norms; for instance, by failing to accord members of certain groups basic human rights.

Finally, although the phenomenon of hate generally seems more serious as you rise up the pyramid, that's not necessarily true all the time. If you are bullied every day at school or experience small acts of bias in the workplace, that might feel worse to you than a single act of systemic discrimination, like being denied a bank loan because of your ethnicity. Context matters, as does the lived experience of individuals.

Ultimately, we should go beyond the pyramid itself and understand hate as a complex *system* of oppression. Multiple levels of the pyramid might emerge in a given society at the same time, with the levels influencing and reinforcing one another. Certainly this is the case in the United States and elsewhere; in recent years we've found the bottom four levels of the pyramid increasingly in evidence.

Because small acts of hate are part of the system, we must take steps to address them, not just wait for mass acts of violence to occur. And as difficult as it might be to pin down causality, the fact remains that we *can* observe relationships between hateful rhetoric and seemingly minor expressions of hate and actual violence.

At his campaign rallies, President Trump routinely disparaged immigrants, Muslims, and other minority communities. Analyzing ADL data, researchers found that counties in the United States "that had hosted a 2016 Trump campaign rally saw a 226 percent increase in reported hate crimes over comparable counties that did not host such a rally." While it can be difficult to ascertain causality, the correlation is striking.

Consider the recent scourge of hate crimes experienced by members of the Asian-American and Pacific Islander (AAPI) community. Bias against AAPIs long has existed in American society, flaring up during periods of social tension, economic stress, or military conflict between the U.S. and Asian countries. But beginning in 2020, some politicians, including Trump, blamed China and Chinese people for bringing the COVID pandemic to American shores. Rhetoric soon gave way to actions.

One study logged 3,800 hate crimes against AAPIs over a roughly one-year period in 2020–2021, up from 2,600 the previous year. And while the causality again can be difficult to pin down, an ADL analysis found an 85 percent spike in hateful tweets directed against people from the AAPI community after Trump tweeted about his COVID-19 diagnosis in early October 2020. These tweets directly slandered Asian people and often referred back to words that the president himself had uttered earlier in the year.

AAPI activists harbor little doubt about the link between anti-immigrant and anti-Chinese rhetoric and hate crimes. As one argued,

"There's a clear correlation between President Trump's incendiary comments, his insistence on using the term 'Chinese virus' and the subsequent hate speech spread on social media and the hate violence directed towards us. It gives people license to attack us."

As of this writing, we seem to be moving even higher up the pyramid. On March 16, 2021, a man killed eight people at three Atlanta-area massage parlors. Six of them were Asian-American women, suggesting that ethnic-based hate was a motive. Although we can't identify the profusion of hateful speech and actions as a "cause" of this atrocity, it seems clear that a broader climate at least contributed to it by making mass violence more likely.

The pyramid also prompts us to conceive of a relationship between the explosion of antisemitic rhetoric in recent years and hate crimes directed at Jews. As I've noted, we're currently seeing record levels of hate crimes. But we're also seeing record levels of white-supremacist propaganda. In 2017, the ADL logged fewer than 500 incidents of white-supremacist propaganda. In 2020, we tracked over 5,000. And this is to say nothing of the actions that Trump and other political leaders have taken to spread and legitimize white-supremacist ideas, whether that entailed credentialing white-supremacist media from the White House Press Office or repeating their rhetoric in official statements.

Although substantively different from the president's proclamations, rhetoric from parts of the Far Left dehumanizes wide swaths of the Jewish population. Radical anti-Israel activists variously label Zionists, Israel supporters, and Israelis as inherently Nazis, "garbage people," or deserving of death. When members of Congress repeat wild claims, such as Democratic representative Betty McCollum did when she accused Israel of apartheid, it serves to enrage and inflame the broader population even if numerous experts dispute this characterization as "slander." Such claims feed outrageous headlines like "Not One Dollar for Zionist Genocide!," an op-ed in the American Communist newspaper *Workers World* that went on to describe Israel as "the Zionist settler state." Wild rhetoric on any side of the political spectrum that is continually pounded into the public consciousness inevitably has an impact.

We might not be able to link hateful ideas causally to a given act of violence, but it seems intentionally naive at best, an outright fiction at worst, to deny any connection. Given enough time, seemingly small expressions of hate will spread. Biases will harden and become reality. And an environment will emerge in which violence becomes possible, even likely.

## Where There's Fire, There's Smoke

That is, unless we do something about it. Jinnie Spiegler, the ADL's director of Curriculum and Training, develops educational materials for use in schools, workplaces, and elsewhere; these tools help to fight hate by disseminating the concept of the Pyramid of Hate and other information. As she notes, the pyramid is useful primarily as a call to action.

The vast majority of people (hopefully) will never encounter a violent extremist group, let alone play a direct role in fighting a tyrannical government bent on slaughtering its citizens. But people *can* and must intercede at lower levels of the pyramid to help reduce the chances that society will see spikes at the upper levels. "We can learn about and counter our own biases," Spiegler says, "interrupt when we hear biased language, and take action to address systemic discrimination in the community through political action, volunteering, education, and more."

It's not difficult to ignore the lower levels of the pyramid, pretend they don't exist, or imagine they aren't relevant to you in your daily life. As Spiegler notes, community leaders and educators are often so shocked in the face of brutal acts of hate-inspired violence that they disavow the presence of hate in their communities. "This is not us," they'll say. While they'll disavow a particular act of hate, they won't really deal with the three levels of the pyramid below it. "We're always pushing school leaders and others to deal with these levels, telling them those incidents don't happen in a vacuum. What I always say in these situations is where there's fire, there's smoke. The fire is the bias-motivated violence, but there's all these smoking embers underneath that and we need to address them."

If you're disturbed by the bias and hate you see throughout society, you can make a difference by putting out the smoking embers around you, as relatively innocuous as they might seem. In Julian Edelman's hands, Meyers Leonard's slur became a valuable teaching moment, one that played out before a mass audience on social media. If just one vulnerable youth read Edelman's posting, became more aware of unconscious biases, and stopped using antisemitic language, that means something. Our discourse just got more civil, our society safer and more humane.

Social media can be a venue for spreading hate, but it can also serve as a valuable means of countering it. I like to think of social media as akin to the human immune system in a way. When someone introduces the virus of hate, others rush in like white blood cells to target the virus, neutralize it, and prevent it from spreading. That's what the second half of this book is about: showing you how you can serve as part of society's immune system against hate in your personal life, at work, and in other important contexts. First, though, let's spend a bit more time unpacking the phenomenon of hate and understanding the dark places where it might lead should we fail to step up and counter it.

# 3

# THE TOP OF THE PYRAMID

A fourteen-year-old girl, scared and traveling without her parents or siblings, steps off a plane in a strange airport. She's never been in this country or on this continent before, never been away from her parents. But here she is, dazed and weary and frightened after her long flight, surrounded by people speaking a foreign language she doesn't understand.

With nobody to guide her, she follows the other passengers to Customs. She reaches into her jacket, takes out her passport, and clutches it tightly by her side. Despite her age, she is trying her best to look calm and assured.

This isn't a school vacation. It's not an ordinary trip to visit family members. Upon departing, she had hugged her parents tight because she had no idea if or when she would see them again. If she gets through Customs, she'll meet up with relatives and make a new life for herself in this country. Although her parents managed to sneak her out, they had no way of leaving themselves and didn't know if they ever would. For now, she would have to manage without them, a refugee in a foreign land.

This girl is now my wife, Marjan. The year was 1985, six years after a revolution had swept away Shah Mohammad Reza Pahlavi and brought an Islamic fundamentalist government to power. Jews like Marjan and her family had lived in what is today called Iran for more than two millennia, settling there following the Babylonian sacking of Jerusalem in the sixth century BC. They had coexisted in relative peace with the majority Zoroastrian and, later, Shia Muslim neighbors, though with the arrival of Islam, the Jews did become second-class citizens. Aside from institutional discrimination, forced conversions, and the occasional pogrom, life wasn't bad.

When the shah's government fell in the early weeks of 1979, members of the Jewish community were on edge. Their worst fears were confirmed when revolutionaries executed arguably Iran's most high-profile Jewish business figure, Habib Elghanian, in May 1979 after falsely convicting him of spying for Israel. The murder sent an unmistakable message to the Jewish community: *You are not safe.* Shortly afterward, a small group of Iranian Jews met with Ayatollah Khomeini; he offered reassurances that the new regime wouldn't target its Jewish population. That was a lie. The government continued to demonstrate its brutality, executing prominent members of the Jewish community such as Gorgi Lavi, Simon Farzami, Albert Danielpour, and many others. As a result, the vast majority of Iran's eighty thousand Jews left the country.

Many Jews went to America or Europe; others fled to Israel. Those who remained in Iran tended to be professionals, small shopkeepers, and other members of the middle class. My future father-in-law, who co-owned a small men's store, was one of these people. Although he felt tempted to leave when relatives and friends abandoned their homeland, his exit options were limited and the thought of starting over as a middle-aged man in a new country seemed daunting. The situation appeared grim, given the rigid, fundamentalist view of Islam espoused by the revolutionaries, but he held out some hope that it might improve.

Unfortunately, it did not.

After the shah's fall, the revolutionary clerics moved to consolidate their power and eliminate their perceived enemies. They conducted a purge in their effort to "Islamize" the country, imprisoning and executing not only former high-level officials who had worked for the shah but anyone who might pose a threat to a new Islamic regime: business executives, labor leaders, intellectuals, artists, journalists, and potential dissidents of any sort. By the fall of 1979, revolutionary courts under Ayatollah Sadeq Khalkhali, known as the "hanging judge," had murdered some 550 people. A terrible chill fell over the country.

Strict religious law was imposed, and a nation that once had pursued modernization and openness fell under the dark, dictatorial shadow of fundamentalist Islam. A system of pernicious, institutionalized gender discrimination took hold. Women and girls were forced to wear the veil regardless of their religion. They were segregated from men at work, at public events, and elsewhere across society. More repressive policies followed. Schools were forced to adopt Islamic curricula. Young children, including Marjan, were compelled to study the Koran even if they were not Muslim. Universities were closed. Alcohol and music were banned. Dissenting media was shut down.

My father-in-law watched these changes play out. One day, members of the feared Revolutionary Guard came into his store, grabbed an employee, pinned him down, and cut his long hair, claiming that it was un-Islamic. The following year, when war with Iraq broke out, life got even worse. Food was rationed, children were drafted into the military, and evenings were spent in the cellars of homes or apartment buildings to avoid bombardments by the Iraqi air force.

By then, my future in-laws desperately wanted to leave, but they couldn't. Khomeini had a raw, unhinged hatred of Israel. It was core to his revolutionary creed; he hoped Shiite Iran would lead its Sunni cousins across the Muslim world in a religious holy war. The new Islamic republic demonized Israel as the "Zionist state" and publicly called for its destruction, an effective way to distract from its own deep shortcomings at home. Alongside this seething hostility, the government surveilled its own Jewish citizens and prevented them from leaving the country. It was a policy seemingly modeled on those of the despotic Soviet Union, which also refused to allow Jewish emigration.

And yet, my future in-laws were lucky—at great personal risk, they eventually managed to sneak their two daughters out of the country. So as not to arouse suspicion, their daughters left one at a time, their departures separated by almost a year. In 1987, my future in-laws themselves managed to get out under the cover of night. In 1988, they settled in Los Angeles with Marjan. They had escaped the revolution, but just barely.

Their experience was terrifying but not entirely surprising. Such stories of escape and exile have marked Jewish experience across countries and cultures. Again, my in-laws were lucky, but their story stands as a sober symbol of the vulnerability of Jews and other minorities. It's also a reminder that cataclysms don't happen all at once. They unfold gradually, almost imperceptibly, over time. A few people might spot the trends early, but many don't. The unimaginable is precisely that, unimaginable—until it isn't. But when the outlines of the horror become clear, it might be too late to escape.

Genocides start like this. They become possible when an underlying social context of hate arises and solidifies over time. From inside that context, hate seems normal and not especially dangerous. Someone shouts a slur at you or spits at you on the street, or they refuse to serve you at a restaurant, or they break off a friendship with you. But at least they're not knocking on your door, pulling you out of your bed, and sending you to die in a concentration camp. At least citizen-led militias aren't taking to the streets and machine-gunning perceived enemies in broad daylight with impunity. Life might be unpleasant, but it's not intolerable. Mass murder remains unthinkable.

And then, one day, the unthinkable happens.

## The Failure of "Never Again"

After the Holocaust, Jews and the world at large said, "Never again." Never again would humanity allow the Jewish people to be marked for extinction. And never again would evil people be allowed to perpetuate genocide against *any* group.

But not five decades after Allied forces extinguished the crematoria of Auschwitz, genocide returned to Europe, this time as part of a war in the former Yugoslavia fought along ethno-nationalist lines. It's a story that happened in the recent past, but too few today remember it.

The conflict between Serbs, Croats, and Muslims in the former Yugoslav republic, while hardly inevitable, did draw on lingering animosities. During World War II, a fascist Nazi-aligned government in Croatia killed hundreds of thousands of Serbs. In Bosnia and Herzegovina (one of the six republics in the Socialist Federal Republic of Yugoslavia created in 1945), local populations acted opportunistically to collaborate with or resist occupying forces, decisions that at times put Muslims and Serbs on opposite sides.

For decades afterward, Yugoslavia's Communist central government subsumed ethnic conflicts through a doctrine of "Brotherhood and Unity." In urban areas, a norm of multiculturalism had taken hold, coming to define everyday life in much of the country.

Adna Karamehic-Oates, director of the Center for Bosnian Studies at Fontbonne University in St. Louis, grew up in Bosnia during the 1980s. Her town, she recalls, had a mosque, an Eastern Orthodox church, a Catholic church, and a Catholic monastery. Although people had a sense of their own identities, ethnic differences didn't influence social life all that much. Bosnian Serbs, Croats, and Muslims celebrated one another's religious holidays and even intermarried. Her father was the town's mayor, and although he was a Muslim, he was good friends with both Serbs and Croats. Karamehic-Oates herself went to school with kids of both ethnicities.

And yet, ethnic grievances and a sense of difference were never extinguished, particularly in the rural areas. They became increasingly salient during the 1980s following the death of Josip Broz Tito, the strongman who had served as Yugoslavia's president. As the economy fell into crisis, the country grew restive. Nationalist ideology flourished, most notably among Serbs (the largest of the ethnic groups) but also among Muslims and Croats. Observers have laid much of the blame for resurgent nationalism at the feet of Serbian leader Slobodan Milošević, regarding him as a "malign" opportunist who sought to ride to power by stoking tribal sentiments among his fellow Serbs.

A pivotal moment came in 1987 when Milošević visited Kosovo, an autonomous province of the Yugoslav republic that saw rising tensions between ethnic Serbs, who tended to be Orthodox Christian, and a largely Muslim Albanian population. A violent mob of Serbs had assembled, and local authorities allowed Milošević to speak, thinking he would help lower the temperature. Instead, he gave a fiery speech that encouraged the crowd and established himself as a leading mouthpiece for Serb nationalism. "You should stay here," he told them. "This is your land. These are your houses. Your meadows and gardens. Your memories. You shouldn't abandon your land just because it's difficult to live, because you are pressured by injustice and degradation." One phrase of his from that occasion — "no one should dare to beat you" — became a powerful rallying cry for Serb aspirations.

Over the next few years, the political situation in Kosovo, Bosnia, Croatia, and other parts of the Yugoslav republic deteriorated, fed by politicians — most notably on the Serbian side — who ratcheted up nationalist fervor. Wars erupted after Croatia and another former republic, Slovenia, seceded from Yugoslavia. And then in 1992, war erupted in Bosnia and Herzegovina, Yugoslavia's most ethnically mixed repub-

lic, with Muslims (44 percent of the population) outnumbering Serbs
(31 percent) and Croats (17 to 18 percent). Bosnia, with its Muslim plu-
rality, declared its independence from the Serb-dominated federation,
prompting the Bosnian Serbs to launch "a military campaign to secure
coveted territory and 'cleanse' Bosnia of its Muslim civilian population."

The war lasted three years and saw the killing of one hundred thousand people, four-fifths of whom were Muslim. It also saw horrifying
acts of genocide against the Muslims; in the town of Srebrenica, Serbs
massacred about eight thousand men and boys and buried them in mass
graves. Across Bosnia, Serb forces "ethnically cleansed" villages of their
Muslim populations, killing some civilians and forcing others to flee as
refugees. "Just like the Holocaust 50 years before," wrote the editors
of a volume of testimonials from Srebrenica survivors, "the violence in
former Yugoslavia in the 1990s was intended to destroy a specific group
of people, their history, religion, and culture."

To at least some of the victims, this genocide seemed incomprehen-
sible. "I didn't do anything wrong to anybody," said one man who had
fled his village after it came under Serbian bombardment and later survived a mass killing near Srebrenica. "As a normal person, you couldn't
have imagined what would happen. How could you presume that such
horrible things could happen?"

As Adna Karamehic-Oates confirms, this is a fairly typical reaction.
As part of her doctoral dissertation and her ongoing work with the Center for Bosnian Studies, she interviewed and researched the accounts of
hundreds of survivors. She found that because of the previous reality of
peaceful coexistence between ethnic groups, the sudden departure from
that tradition "was extremely shocking. When it came, people were in
denial, and they were also unprepared to fight back."

The ability of human beings to cause extreme suffering to fellow humans might elude explanation and occasion a certain amount of shock. But that doesn't mean we can't gain valuable insight from the attempted destruction of a people. Genocide has a cultural basis to it. Politicians arguing for a Greater Serbia didn't simply glorify Serb identity; they demonized and dehumanized Muslims as Other. In 1989, at a religiously charged speech in Kosovo commemorating the defeat of Serbian forces at the hands of the Turks six hundred years earlier, Milošević, according to one account, "spoke of battles in the past and battles to come and praised Serbia as the defender of Christian Europe against Islam."

The event evoked a key theme in Serbian nationalism, the notion that Bosnian Muslims were "Turks" and "forever alien"; indeed, "parasites who lived on the blood drained from the Serbian people." Serbian cultural and religious leaders, including those in the Serbian Orthodox Church, had over the previous few years publicly claimed that Muslims were plotting to ethnically cleanse Serbs from Kosovo. That, combined with the Serbs' memories of what had happened to them during World War II, led to "an intense mass psychology that was to be harnessed and used to motivate and justify genocide in Bosnia, often against those who were former neighbors, friends, in-laws, colleagues, and lovers."

During the late 1980s and early 1990s, the tightly controlled Serbian media portrayed Muslims as evildoers who were plotting to defile Serbian women and subjugate the population. Serbian nationalists also claimed that Muslims were "genetically deformed" in ways that impaired their abilities to reason. Far from a fringe belief, it was advanced by Biljana Plavšić, a prominent scientist and member of the Bosnia and Herzegovina Academy of Arts and Sciences who later served time in prison for war crimes. Shockingly, many respected intellectuals quickly parted with norms of reasoned discourse and embraced a radical Serbian nationalism that disparaged the Muslim Other.

Serbian television reminded viewers of the ethnic group's historic suffering centuries earlier at the hands of the Turks, whom the media implicitly identified with contemporary Muslims. Viewers in Serbia learned that "Jihadi warriors" were threatening ethnic Serbians elsewhere in Yugoslavia and that they would require military protection. As Adna Karamehic-Oates relates, during the war in Croatia, Bosnian Serbs heard sordid tales of what the Croats were doing to the Serbs. "The message was: 'Now they're coming for you in Bosnia.' All that propaganda on the television and radio, it was very difficult in rural places in particular to distinguish between reality and propaganda."

There was, in the words of one observer, "an almost unshakable image of Serbians as the victims of unprovoked foreign aggression." In pleading guilty to war crimes later, Plavšić acknowledged that there had been "a blinding fear that led to an obsession, especially for those of us for whom the Second World War was a living memory, that Serbs would never again allow themselves to become victims. In this, we in the leadership violated the most basic duty of every human being, the duty to restrain oneself and to respect the human dignity of others."

## "Everyone I Thought I Knew I Didn't Know Anymore"

What did it feel like to live through such mounting hate? Eyewitnesses recall a rapid coarsening of everyday life, an eroding of civil discourse that they were powerless to stop. In his memoir *The Bosnia List,* Kenan Trebincevic describes a process of polarization and dehumanization that unfolded over a few years in the Bosnian town of Brcko. As a child growing up in a Muslim family, Trebincevic had interacted freely with Serbs at school and in his neighborhood. But little by little, his world changed.

First, the Serb karate coach he revered began ignoring him after making "weekend warrior" trips to fight with his fellow Serbs in Croatia.

Political discourse became visibly polarized along ethnic lines. Serbian kids in the neighborhood started behaving disrespectfully; as Trebincevic reports, they "stopped talking to me and stared as I walked by. They snickered and called me 'Bosniak' behind my back."

The situation worsened in March 1992, when Bosnia declared independence from Yugoslavia, a move that threatened the region's Serbs (Muslims outnumbered Serbs in Bosnia; the Serbs feared being subjugated by them) as well as nationalist Serbs' hopes to increase territory in the former Yugoslavia for a "greater Serbia." Kenan's Serbian friends applauded the killing of Muslims by Serb paramilitaries right in front of Kenan. "It felt as if someone had pressed a button that turned the world lonelier, like everyone I thought I knew I didn't know anymore."

Not long afterward, Serb soldiers arrived in Kenan's town, and Kenan's Serb neighbors welcomed it; one of them said, "Soon Brcko will be rid of all the Croats and your people too." A grocery-store clerk who previously had been friendly told Kenan not to come back, saying, "You Turks won't be needing to eat much longer." The slur *Turk* puzzled Kenan, as did the sudden shift in how Serbs in his town were treating him and his family. "Neighbors and friends we lived close to our whole lives were abandoning us like rats scurrying from a shipwreck."

Even more shocking, once, after Kenan was shot at, he approached his teacher on the street, seeking assistance. Instead of helping, the teacher, a Serb, put his AK-47 to Kenan's head and tried to kill him. Fortunately, his gun jammed, giving Kenan a chance to run away.

It would not be the last time a former friend or acquaintance tried to harm Kenan or his family members. But by this point, it was too late to escape easily. His father had underestimated the danger. Now, as my in-laws had a decade earlier in Iran, the family would have to survive as best they could, as the exits appeared to be locked shut.

## Closer Than You Think

Pondering the Pyramid of Hate, you might find it plausible enough that prejudices and nonviolent acts of enmity, like bullying or insensitive comments, could lead to small-scale acts of violence, like the desecration of a cemetery or a physical assault on the street. But full-on genocide? Doesn't that sound a little extreme or alarmist?

It might sound that way, and in fact your community might not be on the brink of a human catastrophe. In general, you would be ill-advised to think of the pyramid as an oracle from which definitive conclusions about society's prospects can be drawn. The pyramid isn't a scientifically validated model, and in any case, sociological constructs don't function with the precision of algebraic proofs, nor do they lend themselves to the scientific method as well as, say, Darwin's theory of evolution does.

That said, the path to genocide might be shorter than you think. The pyramid overlaps with other theories about possible warning signs of genocide. One academic expert cites the presence of "hate propaganda" in the media and elsewhere and the presence of "unjust discriminatory legislation and related measures" as warning signs of genocide. Both are reflected on the Pyramid of Hate. Another model, known as "the Ten Stages of Genocide," lists discrimination, dehumanization, and polarization as steps on the path to genocide. Although this model cites other important administrative and political steps that prefigure outbreaks of mass slaughter — such as the organization of armed groups bent on carrying out the atrocities, the rise to power of extremists, and the detailed planning and preparation of genocidal acts — clearly, the prevalence of hate speech and hateful actions of increasing intensity and virulence should set off societal alarm bells.

Genocide is a depressing subject — most of us don't like to think

about it. We prefer to focus on the positive, seeing humanity as fundamentally good and incapable of brutality. But of course, genocide has
happened since the Holocaust. There are recent examples even beyond
Bosnia in other regions of the world—think Rwanda, Cambodia,
Syria. If we're not alert, if we don't expand our sense of the unthinkable, we increase the odds that it can happen again.

The embers of the Bosnian genocide remain very much alive in the
contemporary West. In chapter 7, we'll discuss the "great replacement
theory" that animates the Far Right in the United States, Western Europe, and elsewhere—the idea that dark Others seek to displace the
white population, who therefore must fight back to ensure their survival. Guess where that theory comes from? The Balkans and, in particular, the idea of the Muslim population as usurpers of Serbians' rightful
place.

Right-wing thinkers across the West explicitly reference the Serbian
genocide as a model for them to follow. As one analyst wrote, "The
Bosnian Genocide is a rhetorical and conceptual pillar of the Western
Far Right, an example of the kinds of regimes and policies they embrace and aspire to replicate." Such enthusiasm has already cost lives.
The perpetrator of the Christchurch, New Zealand, massacre of Muslim worshippers listened to a Serb nationalist song as he headed to his
intended killing fields. He had also spent time in the Balkans, visiting
sites where, centuries earlier, Christians had fought with the Ottoman
Turks. His stated goal, expressed in a manifesto, was to exact revenge
for Muslim attacks in Europe.

American extremists, too, are finding inspiration in the Serbs' genocide of Bosnian Muslims. A survivor of the genocide, Columbia University professor Amra Sabic-El-Rayess, described her shock on spotting on Twitter a photograph of a young, armed American extremist
wearing a Serb nationalist insignia. "From where I sit," she wrote, "the
US has moved through most of the steps of the societal disintegration

that I witnessed in the former Yugoslavia: the emergence of the ideo-logical narrative of one group's superiority over the rest; the rise of a political leader who champions that narrative; the leveraging of media to build the alternative truths to morally justify killings of the subju-gated groups; the spreading of hatred and escalation of racial violence that has even divided families at dinner tables."

Sabic-El-Rayess has a warning for Americans: Hate is leading them to dark places, faster than they might think. Genocide in her home country had once seemed unthinkable to her too. She had underes-timated the confoundingly powerful force that dehumanization can exert. Americans must learn the lessons that survivors like Sabic-El-Rayess wish to teach us. We must heed their warnings. And we must take action to counter hate at all levels of the pyramid, whenever and wherever we encounter it.

If these stories still don't convince you, other voices just might. Former extremists I've met have powerful stories to tell. Listening to them, I've learned just how easy it is for individuals — even intelligent, well-meaning ones — to become ensnared in extremist ideology and commit previously unthinkable acts of violence. Once again, I come away with a sense of our civil society's surprising vulnerability. Even if organized mass genocide isn't on our doorstep, and I desperately hope it isn't, broad civil unrest, sporadic clashes, and widespread tribal violence no longer seems unimaginable. Hate is insidious. It's growing. And it's terrifying in its destructive potential. We can't wait for others to stop it. We must stop it ourselves while we still can.

# 4

# THE MAKING OF AN EXTREMIST

Around one in the morning on June 10, 1990, a white Crown Victoria approached the West End Synagogue in Nashville, Tennessee. It was dead quiet; nobody was around. Without provocation, someone sitting in the passenger seat shot about half a dozen rounds with an automatic weapon, shattering one of the synagogue's windows. The car drove off into the night, its headlights darkened to avoid detection.

Nobody was hurt—the building was empty at the time. But the synagogue's violent desecration was unnerving, part of a surge of extremist activity taking place in the local area. Months earlier, someone had drawn swastikas on the building along with the words *Jews Suck* and *Get Out*. Remembering the events decades later, the congregation's former rabbi told a reporter that "[the shooting made] it clear that there were Nazis and Ku Klux Klan members who were really intent on their hatred of Jews, that they wanted Jews dead."

The FBI soon caught up with the triggerman. Leonard William Armstrong was indeed a Ku Klux Klan member; in fact, a fairly important one—grand dragon of the Tennessee White Knights of the KKK. In 1992, he pleaded guilty to the shooting and served about three and a half years in prison. He later claimed to have renounced white supremacism, and he took responsibility for his role that night.

The other person in the car — the driver — was a seventeen-year-old skinhead by the name of Damien Patton. A native of Los Angeles, he had come to Tennessee weeks earlier at the invitation of Jonathan David Brown, a well-known music producer and white-identity preacher who bankrolled white-supremacist groups. Brown put Patton up in an apartment, paid for his living expenses, and introduced him to other local white supremacists as an up-and-coming leader. Patton began attending local white-supremacist events, and in fact, hours before the shooting, he was among hundreds of people who attended a meeting of the Aryan Nations.

Despite his youth, Patton had an insider's knowledge of white-supremacist and antisemitic ideology. This became evident in 1991 and 1992, when he testified about the synagogue shooting to a grand jury. "Jews . . . were felt to be the evil of all problems," he said then, describing the beliefs of white supremacists. "We refer to them as Kikes or ZOG, which stands for Zionist Occupied Government, which means our government was occupied by Jews and Jews only."

Although he could talk the talk and although his role in the shooting proved he was ready to walk the walk, Patton harbored some secrets that he'd kept from Brown, Armstrong, and their friends and that he hoped they'd never uncover.

Damien Patton was living a lie.

Although he'd been a skinhead in California, he previously had hung around with Latino gang members and had developed a close childhood friendship with a Mexican immigrant boy and his family. These facts were bad enough from the Klan's point of view, but there was something else, a part of Patton's background that, if it were ever unearthed, would almost certainly have put his life in danger.

Damien Patton was Jewish.

## A Tumultuous Childhood

Scholars have put forth any number of theories as to what causes people to become violent extremists. Some psychologists theorize that people gravitate toward violent extremism because of a burning need to feel significant. All of us want to be "someone," and some people feel insignificant thanks to specific situations that have arisen in their lives. Extremism seems to inject a sense of meaning and purpose into their reality. It offers the allure of a distinct narrative—a cause in which to believe—and a group of insiders who subscribe to it.

Relatedly, psychologists have suggested that individuals become vulnerable when their ties with social institutions (family, school, work, and so on) fray. In these situations, they might fall under the sway of unsavory influences, developing new, potentially destructive ways of making sense of the world and learning "vocabularies of violence." Other experts have seemed to affirm the role played by social dislocation and feelings such as loneliness and insecurity. The FBI, for instance, includes "feeling alone or lacking meaning and purpose in life," "being emotionally upset after a stressful event," "not feeling valued or appreciated by society," and "believing they have limited chances to succeed" as among the factors that make people vulnerable to radicalization. Research has suggested that extremists have higher rates of childhood trauma than members of the general population.

When it comes to right-wing extremism in particular, scholars have identified five types of people who tend to join up. Some are ideologues who perceive a threat from the Other—Jews, Muslims, and so on—and want to do something about it. Some join for the sense of solidarity, protection, and friendship a group offers. Some want the rush that comes with proximity to violence. Some are angry and inclined to vio-

lence because of traumas in their lives. And some grew up in extremist
settings and were socialized to participate.

I'm no psychologist, so I can't explain with any scientific precision
what exactly prompted a teenager back in 1990 to travel halfway across
the country to engage in white-supremacist activity, much less a Jewish
kid from Los Angeles. What I can tell you is that some aspects of these
theories certainly seem to figure in the story of Damien's descent into
white supremacy.

"Life sucked," Damien says when asked about his experiences growing up during the late 1970s and '80s in the Los Angeles neighborhood
of Torrance. His parents divorced under unpleasant circumstances
when he was about five years old. Although his mother had custody of
the children at first, she struggled to take care of him, his brother, and
his sister on a hairstylist's income, so Damien and his brother went to
live with their dad and stepmother.

Jewish by birth but nonpracticing, Damien's mother spoke Yiddish
on occasion with Damien and took him to bat mitzvahs and other family religious events. Damien was strongly attached to her and hated
being away from her—it was "probably the hardest thing ever in my
life." His father and stepmother both struggled with personal issues,
and their home was a "tough household" filled with tension and animosity.

Damien responded by acting out at school. Once an A student, by
the sixth grade he began to abandon his studies and withdraw socially,
dressing outlandishly in tiger-print bandannas and in general taking his
fashion cues from celebrities like Madonna. After an altercation with a
teacher at the private Lutheran school he attended (she had called him a
fag and he had responded with an obscenity), he transferred to a public
school. There, he continued to rebel by adopting punk fashion, shaving
his head into a mohawk, and wearing jackets with safety pins.

He befriended a Mexican classmate named Pablo, and by the time

Damien was in eighth grade, he was spending long stretches at Pablo's home in Lennox, a gang-infested section of Los Angeles near LAX airport. Warfare between the Crips and the Bloods was then in high gear, and Damien found himself in the frightening and dangerous world of lethal fights and drive-by shootings. He became sensitized to the racial differences that defined the identities of the various street gangs contending for territory and power—not just Black gangs, like the Crips and Bloods, but Latino gangs and white skinhead groups.

## Skinhead

Damien's descent into white supremacism took place over an extremely short period—just a few years. As a freshman in high school, he began hanging out with members of Latino gangs. He gained a reputation as a troublemaker, getting into fights and acting out, and he didn't make it through the year because school officials asked him to leave. He began his sophomore year at a different school and was asked to leave again. By this point, he was living almost entirely without adult supervision. His friend Pablo and his family had moved to a neighborhood located about an hour and a half away, and without asking anyone, Damien would regularly take a Greyhound bus out there by himself. He was beginning to live in abandoned buildings off Hollywood Boulevard, becoming part of an underworld of runaways and drug addicts who had come from across America (incredibly, Damien never actually did drugs himself).

Pablo joined a Latino gang whose members were stealing cars, manufacturing methamphetamine, and committing other serious crimes. Damien began flying gang colors as well; as he tells it, it was "just a thing" that kids on the street did. But his dabbling in gang culture got more serious—and dangerous. Older gang members, hardened criminals with a propensity for violence, came by to hang out with Pablo and Damien. "I think this was my first experience being around a gun," Damien recalls. "It was these Latino gang members getting ready to get into it and have a shootout with the Crips and the Bloods."

At one point during his sophomore year, Damien fell into crisis as he realized that he couldn't hang out with Pablo and the Latino gang members any longer. The older gang members regarded Damien as an outsider because of his race. "What's this white kid doing around us?" they would ask Pablo, who increasingly distanced himself from Damien. "I didn't want to be left out, and I was losing everything, my friends, everything." So Damien, a Jewish kid, did something that today appears astonishing: he joined the skinheads.

As Damien remembers, he knew little at the time about white-supremacist ideology. What he cared about was making friends and feeling accepted. As an alienated, homeless white kid, he found that the skinheads filled his needs. And he gravitated toward them for another reason: he knew firsthand that they were capable of protecting him.

On two separate occasions in recent years, skinheads had accosted him and beat him up. It happened once in a public park and another time while he was practicing with his school's football team. "I knew that skinheads were not to be messed with. They were the tough people if you want to be in a street gang."

Of course, not all skinheads were created equal. There were "street punk" skinheads, posers who tried to look tough but who were disorganized and didn't have any core ideology. And then there were white-supremacist skinheads, a real power on the street. "These were the hard-core ones," Damien says. "If you wanted to go up against the Crips and the Bloods or the Hispanics, you had to be with them. People were scared of them."

Damien's entrance into a local skinhead gang came via Matt (a pseudonym), a man whom he looked up to. Matt was a construction worker who owned his own house, drove a yellow Firebird, and had an attractive girlfriend. He seemed to have everything Damien wanted. After be-friending Damien, he introduced him to members of a white-suprem-acist skinhead gang he led. "It clicked real fast," Damien recalls. "All of a sudden, I had these new friends. The camaraderie was so strong. One day it's like you're the weird kid, the odd man out, the kid who dresses funny. Next thing you know, you've got like ten best friends and girl-friends wanting to hang around you. It's like, whoa — magic."

## "How Can You Be a Nazi When You're a Jew?"

Damien was running with Matt and his skinhead gang in 1988, two years before the shooting at the Nashville synagogue. He remembers this because he spent time at Matt's house in April of 1988 on Adolf Hitler's birthday. The number 88 is significant for skinheads. *H* is the eighth letter of the alphabet, so 88 stands for "Heil Hitler." On this particular day, Damien and his new friends held a celebration at Matt's house featuring a performance by a skinhead band. Although Damien had joined the gang only a few months earlier, he was starting to get deeper into the ideology, listening to white-supremacist music, which, as Damien reflects, served as a key means of indoctrination.

One factor driving Damien to become a believer was his sense that he was a social success, an insider, even a force to be reckoned with. Matt had not only welcomed him but positioned him as something of a leader in the group. As a result, people were taking their cues from him. For the first time in his life, he was popular and dating attractive girls. The attention was addicting, as was the sense that he had a "tight group of friends who I felt weren't going to leave me. I was addicted to what I called family, a family unit that had more stability and structure than my own family did because it was so screwed up."

Damien was enthralled by the sense of power that came with belong-ing to a group. On one occasion about a year before going to Tennes-

see, he and a number of his fellow skinheads went on an excursion to
Disneyland. With their shaved heads, black boots, and bomber jackets,
they were a conspicuous presence. Park security quickly took notice.
They made them remove their jackets, telling them that their obvious
identity as skinheads might incite a riot among the African-American
gang members in the park. "That's when I knew we had power. We're
being compared to the Crips and the Bloods. We're on the same level
as them."

A turning point in his indoctrination came around this time when
Damien, for reasons that he cannot remember, visited his mom's house
and brought a fellow skinhead with him so that she could meet him. His
mother had gotten used to seeing him dressed like a skinhead (combat
boots, jeans, bomber jacket, shaved head) but hadn't thought there was
any ideological component to it. But his friend was wearing something
else: a red armband emblazoned with a black swastika.

Seeing that, his mother realized that he had joined a white-supremacist group. She went ballistic, exclaiming to Damien within earshot of
his friend, "How can you be a Nazi when you're a Jew?"

Damien was mortified. His mom had just outed him. If his friend
went back and told others that Damien was Jewish, he'd be booted out
of the group and physically assaulted. So Damien did the only thing he
thought he could: He denied everything. He chose his friends and what
he thought of as his new family over his birth family and his religion.

"I told my mom in front of my friend to go fuck herself," he says.
"'What are you talking about?' I said. 'We're not Jewish. We have nothing to do with Judaism. So why would you say such a stupid thing?'
And I left home and never went back. I completely rebelled at that moment. I went hard and full-on into white nationalism as fast as I could."

Damien embraced a whole new identity, convincing himself that his
old life simply didn't exist. He "packed away" memories of and attach-

ments to his family. All that mattered were the friends he had in the present and the ideological beliefs that bound them together.

Damien became such an avid proponent of white nationalism that he began to recruit other kids to become skinheads. Remember those abandoned buildings where he sometimes lived? This was "the number-one recruiting point where we found new skinheads." The runaways he met there had fallen through the cracks of society and were every bit as vulnerable as he had been. All it took to win them over was befriending them and filling some of their emotional and physical needs. "So many of the kids, especially the girls, were victims of sexual abuse. It was insane. And we felt like protectors. 'Don't worry,' we'd say, 'I'll protect you from your stepfather in Iowa. He won't get you here.' And then you go out and steal her a flight jacket and boots. She's now a skinhead, even though she didn't know about white supremacy five minutes ago."

## Journey to Tennessee

Damien's identity as a skinhead solidified, and he became even more deeply enmeshed in white-supremacist circles. Older members of the Ku Klux Klan and other white-supremacy movements began approaching him to work with them. Recognizing him as intellectually gifted, they asked him to organize groups of skinheads to serve as security at their events. The validation further filled the hole in his life. Not only had he found a new family; now he'd been plucked out of a crowd and told he was special.

At one point, an older Klansman invited Damien to a secret meeting of white supremacists in Reseda, a town in California's San Fernando Valley. This was a big deal; none of the other skinheads had been invited. The meeting took place in a room on the second floor of a shopping center. Inside, Damien found himself among dozens of hardened white supremacists. These weren't the usual high-school kids Damien hung out with. Many were middle-aged, and some were as old as seventy. They passed out literature to one another, gave speeches, and watched the notoriously racist film *Birth of a Nation*.

As the evening drew to a close, a man told Damien that he was headed out to Tennessee. If Damien could drive him (Damien had managed to buy a car at a police auction), he could live with him out there and help organize skinheads. Damien agreed to do it. His local skinhead friends would feel abandoned, but he didn't care. As a skinhead, he had come to idolize the Ku Klux Klan. And now he was going to Tennessee with them. By his way of thinking, he had hit the big time. He was all of seventeen years old.

Upon arriving in Tennessee, he received a warm welcome. His past in California—his Jewish heritage, his friendship with Pablo, his experiences hanging around with Latino gang members—was all erased. Nobody knew about any of it, and Damien wasn't about to share those aspects of his old life. His new friend introduced him as a budding white-supremacist leader, and everyone believed it.

Thanks to his new friend, Damien was soon hanging out with high-level members of the Aryan Nations and their wealthy financial backers. One of them, the well-known music producer Jonathan David Brown, quickly became a kind of father figure to him, paying his expenses and inviting him to spend weekends on a big farm he owned outside of Nashville. Damien's proximity to power and money boosted his popularity among the local skinheads. He was perceived as a kind of golden child, an experience that he found intoxicating.

On the evening of June 9, 1990, Brown introduced him to Leonard William Armstrong, grand dragon of the Tennessee White Knights of the KKK. Brown vouched for Damien, telling Armstrong he was an up-and-coming leader and "our guy." With other skinheads, they went into downtown Nashville and as a group harassed some Black men who happened to drive by; Damien told them that he "had a ticket for them to go back to Africa." Afterward, Armstrong asked Damien to drive him to an undisclosed location. It turned out that they were headed to the West End Synagogue.

In the aftermath of the shooting, the FBI suspected Damien of involvement. Not wanting Damien to speak to the police and potentially implicate a Klan leader, Brown arranged for him to flee Tennessee. He was to live on the road with a girl he was dating, and Brown would pay his expenses. Damien returned to Tennessee toward the end of 1990, and following disputes with Brown and Armstrong, he went to Hawaii to reconcile and live with his father. Over a period of months, he joined the navy, underwent basic training, and was sent to a base in Virginia. At about this time, he also reconciled with his mother and stepfather.

In 1991, while he was living in Virginia, the law finally caught up with him. He was about to deploy to the Persian Gulf for his first of two tours of duty, and his use of his Social Security number as part of the deployment process allowed law enforcement to track him down. As part of a deal he subsequently made, he returned to Tennessee after six months of deployment to testify against Armstrong and Brown. He agreed to serve a period of probation and did so while subsequently redeploying to the Gulf. Armstrong received a sentence of three and a half years in prison, as I've noted above, and Brown received a sentence of over two years plus additional penalties. Since Damien was a minor when the synagogue shooting took place, he was dealt with as a juvenile and his record was sealed.

## Coming to Grips with the Past

Just as Patton had reinvented himself when first traveling to Tennessee, pretending his past didn't exist, he now tried to do something similar, pretending that he had never been a white supremacist. This process was tricky at first. While living in Hawaii and then Virginia, he underwent a months-long process of separating himself from the movement, renouncing his hateful beliefs and symbolically distinguishing himself from skinheads by growing facial hair. Others took notice. On Damien's first day of basic training, a Black drill instructor spotted the white-supremacist tattoos on his body and said, "Well, Robert E. Lee, looks like you and I are going to have some fun together." At graduation, after Damien had finished first in his class, that same drill instructor approached Damien's mother and stepfather and acknowledged what a tremendous transformation he had undergone.

After basic training, however, Damien continued to consort with local white supremacists in Virginia. As he explains it, he was friendly with soldiers of many races, but he feared that the Ku Klux Klan or others in the movement would suspect him of talking to the FBI, so he wanted to give an outward appearance of loyalty. Hanging out with local skinheads was a way of covering his tracks.

After he testified in open court, he was terrified that the Klan would come after him. He was especially frightened because Brown had served as his own attorney and personally cross-examined Damien on the stand, an experience that Damien found extremely intimidating. Fortunately, Damien was transferred to California, and it was there that the true erasure of the past could begin. "I was on a new duty station, nobody knew about my background, nobody knew anything. I was just this smart guy in the military who was advancing quickly."

By the time he was twenty years old, he was completely done with white supremacism. He spent several years in the military. Following his discharge, he pursued a love of auto racing and worked in NASCAR as a mechanic. He then taught himself to code and went on to become an extremely successful entrepreneur. A software company that he founded, Banjo, became a high-growth start-up that, over a period of years, attracted almost a quarter of a billion dollars in financing.

As Banjo's CEO, Damien gave well-received speeches at industry conventions and was touted as a success in the *Wall Street Journal,* the *New York Times,* and elsewhere. In his public appearances, he told an inspiring story about his difficult past, including his homelessness and gang activity, but he omitted any mention of his youthful white supremacism. The only people he told of it were his family and close friends. Nobody ever found out about Damien's sealed juvenile legal record. Indeed, Damien has no legal record as an adult and had several government background checks come back clean. However, as Damien would soon learn, he couldn't outrun his past.

In April 2020, some three decades after the incident at the West End Synagogue, the technology-news website *OneZero* ran an investigative report detailing Damien's youthful white supremacism, complete with a photo from a 1992 newspaper report showing Damien giving a Heil Hitler salute. Other media picked up the story, which took on special relevance due to the somewhat controversial nature of Banjo's product: software that allowed law enforcement agencies to make practical use of digital data from traffic cameras, 911 calls, and public-transportation GPS. (For instance, if a child was abducted, authorities could check traffic cameras to aid in an investigation.) Weeks later, Damien resigned as Banjo's CEO, publicly apologizing for his actions. He withdrew for a year to take stock of his life and explore his past.

And that's how I came to know Damien Patton.

## An Unlikely Friendship

In April 2020, when the news about Damien broke, Jim, a friend of mine in Silicon Valley, texted me and asked if I was available to talk.

These were the early days of the pandemic, so I took his call on a warm spring afternoon while standing outside my house. As he summarized Damien's story, I found myself spellbound. Here was a high-profile tech exec heading one of his portfolio companies, and he had just been outed as a former Nazi skinhead. Oh, and he was Jewish. I was the head of ADL; would I be willing to talk to him?

As is my practice (remember Meyers Leonard), I agreed without hesitation. ADL as an organization has long tried to exemplify the Jewish concept of *teshuvah,* or repentance. Everyone has the ability to atone for misdeeds and seek forgiveness. All of us have the duty to help in that endeavor if we can and not write anybody off out of hand.

I quickly researched Damien online, discovered coverage of the allegations, and followed the trail on social media. I called my general counsel, a longtime ADL employee, and asked if he had heard of this story. He went into the files and, sure enough, found a hard-copy ADL bulletin from 1990 in which we had written about the incident.

I spoke to Damien late that same day. It wasn't an especially long conversation. He told me he had been at the airport in Long Beach when the news broke, returning from a visit to his ailing mother. Damien was glad that our mutual friend had made the connection, but judging from his short, clipped sentences, he seemed to be struggling to come to grips with what was happening. Even in that short conversation, I could hear him wrestling with his guilt not merely about the life he had led so long ago but also about the life that he was leading now and the incomplete picture that board members, employees, customers, and investors had of him.

I offered to speak with Damien again when he returned to Salt Lake City, where he lived with his wife and where Banjo was headquartered. He agreed. A few days later we connected via Zoom. During that session, he rarely looked at the camera. His eyes were glassy and his mind seemed elsewhere. He spoke haltingly as he described his situation. While Damien didn't come out and say it, I sensed that the board was forcing him to separate himself from the company that he had founded. A few weeks earlier, he had been scaling his business and steering it toward an IPO. Now he had crashed to the ground, the wreckage of his life burning around him.

Despite that difficult conversation, Damien and I struck up a friendship and began talking to each other on the phone almost every week —a relationship that, as of this writing, remains ongoing. During the first few months, we spoke mainly about him and his situation; as a former entrepreneur myself, I could relate to his perspective better than many others. From there the topics broadened: life, religion, hate, and, of course, his past and how he might best come to terms with it in the present.

When people learn of my relationship with Damien, many of them ask why I have invested so much time in it. The answer is simple: I genuinely like Damien and believe that his contrition, his repudiation of white-supremacist ideology, and his desire for forgiveness are genuine.

Damien always claimed to me that he didn't know Armstrong intended to fire his weapon, and in a media interview, Armstrong confirmed this. He absolved Damien of any responsibility, saying, "This was not something that had been plotted or planned out. I was a drunken idiot acting spontaneously, and I got him in trouble. He didn't know that I was going to do the thing that I did."

And yet despite this exoneration, Damien has never shrunk from his moral responsibility. He feels that he is culpable for the incident. "I'm not absolving myself," he told a member of my team. "Listen, I was in those hateful groups. I believed in that hateful shit. I was never full-on 'I hate people' because, again, I was there for a sense of belonging. That's why it was as easy for me to get out of it as it was to get into it. Unlike some people, I wasn't raised in a life of hate. I hated my [childhood] situation, but my parents didn't teach me to hate people based on race, religion, whatever."

Interestingly, while Damien had never publicly owned up to his past misdeeds, he had privately been making amends all along. He donated to Jewish causes and encouraged diverse hiring at Banjo. But as we talked about his process of *teshuvah,* he acknowledged that it was time to go deeper and honestly confront everything in his past.

Eager to help, I tracked down the West End Synagogue's religious leader during the early 1990s, Rabbi Ronald Roth, who agreed to speak with Damien. The two have since developed a relationship. Damien and I hope to visit the synagogue together, and he will make amends by giving a talk or performing some other community service. In addition, Damien has reconnected with his Jewish heritage; he's engaging with a local Chabad congregation in Salt Lake City, spending time with a rabbi studying Torah, and using his newfound knowledge of himself to process his past.

It's too early to tell what place Damien will be able to find for himself in the wider world, but so far he has made admirable progress in healing himself internally. He's reinventing himself once again, but this time in a way that is more profound, transformative, and potentially inspiring than before. I can't wait to see what the future holds for him.

## Reining in Cancel Culture

I'm sharing Damien's story with you for a couple of reasons. First, it's important to realize that people who think and behave in hateful ways can often be redeemed, even individuals who join extremist organizations. It's tempting to want to "cancel" such people once and for all, and of course moral censure is important when people don't repent and cling stubbornly to repugnant views.

But we should reserve such censure for the worst culprits, those whose serial offenses over a long period demonstrate a deep refusal to change or even acknowledge an alternative point of view. In contrast, canceling individuals for a single example of poor judgment or a few errant social media posts or even a material transgression made in their youth strikes me as wrong. We should strive to adopt less of a cancel culture, which attacks individuals for any failing, and more of what a friend of mine, entertainer Nick Cannon, describes as a "counsel culture." We should try to work with those who have sinned rather than simply sitting in judgment of them.

When people engage in genuine reflection and express a sincere desire to change, we need to hear them out. We should work with them, knowing that our efforts remain worthwhile even if they ultimately fail. All of us would be wise to remember that we're all human. We all make mistakes.

Individuals aren't the only ones susceptible to being canceled; companies and organizations face calls for such treatment as well. I experienced this when, during the summer of 2020, a coalition of about 150 progressive groups, including civil rights, pro-immigration, anti-Israel, and left-wing radical groups, launched a campaign called Drop the ADL, accusing us of numerous misdeeds dating back to the 1950s.

The vast majority of these claims were ridiculous—factually incorrect and hateful. The campaign claimed that the ADL blacklisted organizations that criticized Israel and accused them of antisemitism. This is pure fiction; we ourselves criticize Israel, and we work with other groups who do so as well. We take great pains to avoid conflating criticism of Israel with antisemitism. Likewise, the campaign claimed that we supported actual antisemites, notably Donald Trump. It's true that we did support Trump's move of the U.S. embassy to Jerusalem (a measure that President Clinton called for back in 1994 and that has had

bipartisan support ever since). But the notion that we support Trump
and that we decline to call him out for bigotry is laughable. Just look
at the public record or, for that matter, this very book. And these are
just a couple of items out of a whole litany of mischaracterizations and
half-truths advanced by Drop the ADL.

The campaign came at us out of the blue, and weirdly enough, the
coalition supporting it included a few organizations that had previously
worked alongside ADL on various initiatives or had joined us in sign-on
letters. I was especially incredulous to see that one of the groups calling
for us to be canceled was an NGO started by a good friend of mine.

When I reached out to him, he informed me that he was no longer
involved with the organization and had no idea why it had come out
against the ADL. He arranged for me to speak with the executive director. During a Zoom call, the executive director opened the conversation by apologizing, noting that he should have contacted us before
signing on to the campaign. As he told me, "I learned long ago that
before you call someone out you have to be willing to call them in."

I had similar conversations with a few other signatories. They shared
with me that they had signed on simply because they were asked to
do so but admitted that they hadn't scrutinized the claims circulated
by the campaign's creators. Because we at ADL started a dialogue and
talked through any misconceptions with these groups, I'm hopeful that
we will deepen our relationships with many of them as a result of this
incident.

As painful as this episode was, it wasn't my first brush with a cancel-culture mentality from so-called progressives. In 2018, two men
sitting in a Philadelphia Starbucks were wrongly arrested, apparently
for no other reason than that they were Black. Shortly after the news
broke, Starbucks contacted me about the possibility of ADL assisting
with antibias training for employees. Given that our organization had
deep expertise and considerable capabilities in this realm, I agreed. Days

later, the company issued a press release announcing a plan to train all their employees in an effort led by several civil rights organizations, including ADL.

This news prompted former Women's March cochairs Tamika Mallory and Linda Sarsour to inject themselves into the conversation and stir up a media frenzy. Both women were controversial for their past actions, which included aligning with the notorious antisemitic demagogue Louis Farrakhan and making extreme, hateful statements about Israel and Zionism. Now the two used social media to blast Starbucks for including ADL in its employee-education effort. Ignoring our long history of advocating for civil rights, Mallory slandered ADL by falsely accusing us of "constantly attacking black and brown people" and calling for a boycott of Starbucks. Sarsour made the outrageous claim that ADL was "an anti-Arab, anti-Palestinian organization that peddles islamophobia and attacks America's prominent Muslim orgs and activists and activists and supports/sponsors US law enforcement agents to travel and get trained by Israeli military." These were flat-out lies, and yet they quickly spread on social media and metastasized across the internet.

Cowed by the explosive controversy, Starbucks quietly took ADL off the list of organizations that would design and implement its antibias training. In the end, we still advised the company, but we were not involved in a meaningful way. Now, I deeply respected that Starbucks needed to use that moment to tackle anti-Black racism head-on. I agree with the premise that Black-led organizations typically are the best choice to lead trainings against anti-Black racism, even with the support of others in the civil rights community, just as I believe that ADL and other Jewish organizations are best prepared to lead efforts to understand and dismantle anti-Jewish hate. Still, I was angry and, as a former Starbucks executive, stung by the experience. As a friend explained, it was "Cancel Culture 101" and evidence that online detractors and anonymous trolls could hold hostage even global corporations committed to doing good.

Another outburst of cancel-culture rejectionism broke out in April 2016 when I accepted an invitation to speak at the annual conference of J Street U, the campus organization of J Street, a liberal Zionist advocacy organization. Many in pro-Israel circles reject J Street because of its liberal positions and frequent criticism of the Israeli government, but I agreed to address their students because I believed it was important to engage them. They, too, are part of the Jewish community, and their liberal Zionist views fall well within the norms of reasonable debate. Frankly, these views align with those of a large majority of Jewish college students in America. Moreover, J Street deeply believes in a two-state solution, a policy that aligns with ADL's long-standing position and my personal view. I fervently believe that Palestinians should be able to realize their national aspirations in a way that provides them with dignity and equality, although any such arrangement must ensure the safety and security of Israel's citizens.

In advance of the anticipated controversy, my staff and I took a series of steps. I wrote an essay explaining why I decided to speak at J Street U. We released my remarks days before my talk. ADL regional offices contacted conservative members of our regional boards to explain my decision before the actual appearance. And yet, as predicted, news of my remarks spawned a series of angry denunciations, outraged op-eds, and furious phone calls from right-wing commentators. Some derided me as anti-Israel, as a self-hating Jew, and so on. I was expecting the epithets and did my best to ride out the storm.

Almost a week after the event, I met a friend for drinks across the street from the ADL's office. We're on the tenth floor of a large skyscraper in midtown Manhattan, sharing space with many tenants and even some UN agencies. Imagine my surprise when, from across the street, I saw a group of well over a hundred protesters converging on the building from multiple directions. I called my chief of staff as I watched the group fill the space. Some of these protesters blocked the doors; others sat down and locked arms in a circle in front of the guard's desk.

On the advice of my security team, I entered the building via a back entrance. When I arrived upstairs, my team gave me a quick appraisal of the situation. When I learned that these protesters were targeting ADL, my first thought was that the conservative criticism had influenced them and that they were fuming about my remarks.

Far from it.

In fact, it was If Not Now, a Far Left organization that specifically targeted ADL. They sought to make the point that, despite my recent appearance at J Street U and the lamentations of some on the political right that ADL was too liberal, we actually *were not* a progressive organization. Rather, they claimed that ADL was part of a nefarious, conservative, pro-Israel establishment. They derided our commitment to Palestinians' dignity and equality as little more than a masquerade.

Some younger members of my team suggested we defuse the situation. I was unsure about the feasibility of this idea, but it seemed worth a shot. We posted a message on social media inviting the group to come upstairs, meet with us, and talk through our differences.

The answer came swiftly: no.

The agitators didn't want to meet with us. This was about posturing for effect, not pushing for actual change. They didn't want dialogue because they literally had nothing to say to us. It was cancel culture through and through.

In truth, none of these episodes should have surprised me. ADL routinely fields intense criticism from activists on both the left and the right. Those on the Far Left claim that ADL's commitment to civil rights—and my personal commitment as ADL's CEO—is weak, compromised by an excessive support for Israel. Detractors on the extreme right regularly contend that we are too beholden to progressives and don't support Israel and Jews enough. These critics often target me personally, claiming that I'm "destroying the ADL" and that I should resign.

I welcome discussion and debate from both sides—it's part of the price of operating in the public domain. And it certainly can help you to see your blind spots and improve your performance. But ADL's hardened critics often don't seek constructive discussion or corrective debate, an honest give-and-take from which both sides might learn. In our poisonous political environment, their intent is to demonize our organization, discredit our work, and treat ADL as an enemy.

One can discern on both the Far Left and the Far Right a certain illiberalism at play, a rejection of dispassionate and self-reflective political discourse and the notion that it can serve as a path to growth and progress. The impulse to cancel others rather than doing the harder, more honest work of engaging with them is a natural outgrowth of this mentality. For the good of our civil society, I deeply believe we should strive to conduct dialogue in good faith even with people who don't share our views. Yes, we should be prepared to ostracize individuals and groups if necessary, but only in extreme situations when their views are clearly hateful and they've shown an unwillingness to learn from their mistakes.

At its worst and most extreme, the impulse to cancel others rather than engage with them can express naked bigotry, potentially leading to violence. Some quarters of the anti-Israel movement have attempted to cancel the entire country of Israel as well as its supporters. A tenet of the Boycott, Divestment, Sanctions (BDS) movement against Israel is "anti-normalization," a position that essentially criminalizes Zionism, the simple belief in Jews' right to self-determination in their ancestral homeland. Anti-normalization advocates nonengagement with Israelis unless they adopt maximalist political positions akin to those of BDS.

Because of their commitment to anti-normalization, some prominent BDS groups in the United States have gone so far as to explicitly bar dialogue with not just pro-Israel groups (which alone would be problematic) but all Israelis based on nothing more than their nationality.

This posture is troubling at every level, since poll after poll has shown that the overwhelming majority of American Jews consider a connection with Israel to be an integral element of their Jewish identities. An organized movement that actively and intentionally seeks to excoriate the entire community in this manner isn't political. It's prejudice, plain and simple. Such delegitimization of a people helps create an environment in which malice and even outright violence toward those very people becomes possible.

In one notable 2018 case, Rabab Ibrahim Abdulhadi, a professor at San Francisco State University, posted on Facebook that she considers "welcoming Zionists to campus . . . to be a declaration of war against Arabs, Muslims, Palestinians and all those who are committed to an indivisible sense of justice on and off campus." Imagine if an individual in a position of authority made such a blanket statement toward any other minority. China's treatment of its indigenous Uyghur population has drawn harsh criticism from human rights activists worldwide, but would a professor ever utter such a statement directed toward Chinese exchange students studying on campus, let alone Chinese-American students? If that happened, would any university administration tolerate such naked bigotry?

Abdulhadi has never expressed regret for her post and in fact has doubled down on her message several times since. If you think that groups and activists on the political left have distanced themselves from her, think again: Abdulhadi continues to be venerated by the most prominent anti-Israel organizations, featured as a guest speaker, and cited glowingly. And this example is just one among many.

In 2018, a leader of a chapter of a prominent national civil rights organization, Zahra Billoo of the Council on American-Islamic Relations (CAIR), tweeted that she "is weary of fellow Muslims who collaborate with or normalize Zionists." Remember, this is a U.S. civil rights leader talking about engagement with fellow citizens.

During an April 2019 student association meeting at UC Berkeley, a student speaker reportedly argued that identifying as Zionist or befriending Zionists renders a person "complicit in the prison-industrial complex and prison militarization and modern-day slavery." It is hard to fathom the connection between private prisons in the United States and the functioning of a small democracy in the Middle East. Such unmitigated and abhorrent slander serves as a kind of kindling for what could become a full-blown inferno.

Certainly the demonization of Zionism and the efforts to equate Zionists with Nazis bodes ill for the physical safety of American Jews. I say this because virtually nobody in any segment of society believes that Nazis have a place in public life. Society does not condemn those who oppose Nazis. If the notion that Zionists are Nazis goes mainstream, given that the overwhelming majority of American Jews believe in Zionism, the danger will be profound. Far from a reasonable and measured response to injustice, cancel culture in this instance serves as a means not of creating a more just society but of perpetuating injustice and marginalizing a vulnerable minority. The inability of some self-proclaimed anti-racist activists to recognize the root evil of anti-Jewish racism is confounding. It is essential that people of good faith disavow such hate in the strongest terms wherever they find it, regardless of their views about the Middle East conflict.

## Understanding the Origins of Hate

In addition to prompting us to reassess cancel culture, Damien's story holds relevance for another reason. Although there is no single path that leads someone to become an extremist, one truth seems clear: The vast majority of hate-group members and extremist sympathizers aren't dangerous psychopaths. Far from it. They're often people who are broken in some way and for whom racist, antisemitic, anti-immigrant, or other hateful ideology comes to fill a void.

Acknowledging this reality doesn't excuse hate. But if we want to try and stop the spread of bigotry, we must seek to understand its causes. While some find meaning in the ideology because it confirms their internal biases, many others, like Damien, find that involvement in hate groups salves a wound that they previously had not recognized. It gives them a sense of belonging, a feeling of power, a sense of meaning, or some other emotional or intellectual satisfaction. Such insights should inform how we deal with people who commit hate crimes. In some cases, severe punishment might be the right answer. Other times, we can use models like restorative justice or community service to rehabilitate attackers, helping them to see that the world is different than what they had imagined.

This analysis doesn't apply solely to right-wing extremists. Scholars of Islamist terrorism have drawn similar conclusions. The British criminologist Andrew Silke has spent time conversing with jihadists in prison. "When I ask them why they got involved, the initial answer is ideology. But if I talk to them about how they got involved, I find out about family fractures, what was happening at school and in their personal lives, employment discrimination, yearnings for revenge for the death toll of Muslims." Ideology matters, but for jihadism and violent extremism in general, complex personal circumstances contribute to the radicalization of individuals, circumstances that might bear some resemblance to Damien's.

Hardened extremists prey on young potential recruits, sniffing out their emotional, psychological, and physical needs and presenting their racist ideology and hateful activities as the answer. They do it so effectively that even a rebellious Jewish kid from California can find himself breaking from his family, adopting antisemitic beliefs, and helping to desecrate a synagogue. Turning a person inside out this way is no small feat, but it's one that happens again and again.

One Austrian researcher who went undercover on various extremist networks puts it this way: "The lowest common denominator was people who were in a moment of crisis. The recruiters did a good job of tailoring their propaganda to pick up vulnerable individuals."

To fight back and prevent our society from scaling the Pyramid of Hate, we must take steps collectively to vaccinate people who might be vulnerable to the virus of hateful beliefs. Before his past became public, Damien provided antibias training to his workforce at Banjo, hoping to ensure that no one at his company would go down the same wrong path that he had. It's a reminder that all of us must do our part, no matter our resources or station in society.

## Our Collective Vulnerability

Ultimately, I believe Damien's story dramatizes how vulnerable society is in the face of hate. Damien's unstable upbringing, sadly, is far from unique. Millions of young people experience physical or sexual abuse, poverty, broken homes, and other traumas that might make them more vulnerable. The vast majority of these people don't go on to join hate groups or become involved in synagogue shootings, but a small percentage of them do, tearing our social fabric by threatening public safety and spreading hate.

(A sad coda: The shooting in which Damien was involved wasn't the last time Nashville's West End Synagogue was targeted. In 2015, someone fired a single shot at the building, leaving a bullet hole but, thankfully, causing no injuries or loss of life. The threat posed by extremists hasn't gone away. In fact, it's worse than ever.)

To prevent societal breakdown, let alone widespread violence and even another genocide, we must prevent *everyone* from traversing this path. And the way to do that is to mobilize all of society to fight back against hateful beliefs before the vulnerable among us can encounter them.

Our task is even more urgent given the powerful tools now in the hands of racists, nativists, and antisemites of every stripe. I'm not talking about weapons of mass destruction—although in a way, I suppose I am. A nuclear bomb can destroy life in a wide swath of territory. Facebook, Twitter, and other social media platforms are weapons of mass disinformation that in a single stroke can infect billions with the virus of hate. These companies serve as super-spreaders of intolerance, achieving reach and scale beyond any previous platform in the history of humanity.

Hate has long posed a threat, but today our democracy hangs in the balance because of the unparalleled ability of extremists to exploit technology to reach vulnerable people and recruit them to their causes. Add in demagogues who deliberately harness traditional and social media to legitimize and spread conspiracy theories and other fake news, and we have a crisis on our hands, one far more consequential and potentially catastrophic than nearly anyone ever could have imagined.

# 5

# HATE BOOSTERS

I magine that a dear friend or family member of yours has passed away, and you're struggling with intense feelings of grief and sadness. Since restrictions on public gatherings remain in place, you can't show up at a memorial in person to say goodbye to your loved one. The only option is to attend an online service on the digital platform Zoom.

You log in at the appointed time; the familiar rectangular Zoom window appears on your computer screen. A few dozen people are already logged in, many family members and friends who, like you, feel emotionally bereft. There is some small talk, and then a religious leader appears on the screen and begins the proceedings. He welcomes the grieving guests and asks everyone to please mute their microphones before starting with a few opening prayers.

Perhaps ten minutes into the service, as the leader transitions into a eulogy for your loved one, something horrible happens. Without warning, one of the guests starts to scream obscenities. The cleric asks him to stop, but now everyone is jolted by unbearably loud heavy-metal music pumped into the session. Another guest's screen changes to a grotesque caricature of a Jewish person.

The host tries to kick out the culprits, but more people log on, bearing photos of Hitler as their IDs. They fill the chat with hateful messages aimed at members of your religious group. They celebrate leaders who perpetrated terrible atrocities against your forebears and claim gleefully that those atrocities never happened. Unable to stop the torrent of filth, the desperate host keeps apologizing and finally clicks End, abruptly terminating the memorial service.

You are left staring at the screen, feeling stunned and helpless. Such messages normally would shock you, but in this context they're inconceivably cruel and heartless, a desecration of the holy space your loved one and the community of mourners deserved. What kind of people do this? Your hands are shaking and your heart is pounding as you close the window on your browser and slap your laptop shut.

Something very similar to this imagined scenario played out in August 2020 when Jewish mourners congregated online to sit shivah (the Jewish practice of paying condolences to the family of a recently deceased person) for a sixty-eight-year-old British woman named Linda Huglin. As participants in the Zoom call later recounted, a handful of agitators infiltrated the session and spewed antisemitic venom. "They were playing films of the Nazis, of Hitler," one said, "they were shouting that there was no Holocaust, and they showed images of people saluting Hitler. It was horrific."

This incident was part of a disgusting wave of antisemitic "Zoom-bombing" that hit the digital world during the opening months of the pandemic, when communities around the globe were first learning how to use Zoom's technology. The ADL tracked numerous such events; anti-Jewish bigots disrupted online funerals, community classes, prayer services, Torah study sessions, school board meetings, Holocaust memorial services. In some of these incidents, antisemites threatened violence, shouting phrases like "Jews in the ovens" and "kill all Jews." Haters also interrupted online university lectures, secondary-school classes, government meetings, and other gatherings, shouting or posting not just antisemitic messages but often racial and xenophobic ones.

Zoom-bombing is hardly the only manifestation of online hate.

Just as harassment is endemic among players of online games, research shows it's widespread on Facebook, Twitter, Instagram, and elsewhere on social media. In 2021, over two-fifths of Americans in an ADL survey reported that they had been victimized online by harassment, and over a quarter had suffered severe harassment, such as threats to their personal safety or stalking. These numbers were steady from the previous year, even though social media platforms claimed in the intervening months that they were aggressively policing themselves. About a third of respondents reported harassment on account of their gender, race, political beliefs, and so on. The effects of hateful messages encountered online are profound. Nearly a quarter of participants in our survey suffered distress such as anxiety or difficulty sleeping.

Online harassment belongs in level 2 of the Pyramid of Hate, which encompasses a range of individual acts of hate, including hate speech. Level 2 is bad enough, but I'm concerned about how emerging models of media can thrust society up the pyramid with dizzying speed. In addition to serving as a venue for inflicting pain on others, social media is a super-spreader of extremism, a powerful tool that ideologues are harnessing to propagate their intolerance, recruit new members, and even organize violent attacks. Far from tamping it down, politicians and influencers exacerbate the problem by echoing and legitimizing hateful conspiracy theories among their vast audiences of followers.

*But certainly,* you say, *the big social media companies are taking swift action to neutralize this threat. After all, they care about their customers and their well-being. They want what's best for society and their shareholders. They have made public promises and are doing everything in their power to ensure that the virus of hate doesn't spread, leading to violence in the real world.*

Don't make me laugh.

With rare exceptions, the responses of Facebook, Twitter, Google, and their ilk have been lackluster and shamefully incomplete at best, criminally negligent at worse. Even now, following the January 6 insurrection, the sickening status quo continues. If these companies had the desire, they could dramatically curtail the presence of hate on their platforms, but again and again, they have opted not to, offering tepid excuses instead of real action. This must change and fast if we are to prevent violence from spreading and society from veering dangerously out of control.

## Life in the Rabbit Hole

Damien Patton became radicalized a generation ago by individuals whom he met in his local community and whom he knew personally. For Caleb Cain, it happened a bit differently. In 2014, when Cain was a college dropout down on his luck, he discovered videos on YouTube of right-wing commentators, agitators, entertainers, and activists. Over the course of two years, he became sucked into what he describes as "an alt-right rabbit hole." "I just kept falling deeper and deeper into this," he said, "and it appealed to me because it made me feel a sense of belonging. I was brainwashed." By the run-up to the 2016 election, he was watching content with overtly white nationalist themes.

Cain's experience is frightfully common. A journalist profiling him noted that he'd "heard countless versions of Mr. Cain's story: an aimless young man — usually white, frequently interested in video games — visits YouTube looking for direction or distraction and is seduced by a community of far-right creators." One study implicated social media in the radicalization of *half* of all members of extremist groups over an eleven-year period ending in 2016.

Cain was fortunate; he found his way out of the rabbit hole. Today, he alerts others to the dangers of right-wing extremism, attracting death threats from extremists in the process. But in other cases, self-radicalized individuals have gone on to cause profound damage. Think of Dylann Roof, the teenager who slaughtered nine African-Americans at the Emanuel African Methodist Episcopal Church in June 2015. And Robert Bowers, perpetrator of the heinous attack on Pittsburgh's Tree of Life synagogue, the most violent antisemitic attack in American history. And Brenton Tarrant, who massacred fifty-one in the Christchurch attacks. Indeed, 90 percent of "lone actors" who committed acts of mass violence in 2016 were radicalized online. Like Roof, an increasing number of such lone wolves plotting or committing violent acts are teenagers, radicalized with increasing speed in the comfort of their living rooms. Some of these youngsters are even leading terrorist cells in hate groups like Feuerkrieg Division or violent fundamentalist organizations like ISIS and Al-Shabab that coalesce online and then strike out in the real world.

We all know at this point that hate is widespread on mainstream social media networks like Facebook and Twitter. Still, the numbers are sobering. ADL researchers analyzed tens of thousands of Facebook posts collected over nine months in 2020. We found that about one in 150 posts contained vitriol toward Blacks, and one in three hundred expressed hatred toward Jews. That might not sound like a lot, but it's enough to open up a wide audience for antisemitic groups and individuals.

Hate is even more prominent on niche social media platforms like 8chan and Gab. With little moderation and virtually no policing of postings on these sites, a toxic loop of hateful ideas and violent action can take hold. My colleague Eileen Hershenov, senior vice president of policy at ADL, has noted how "online propaganda can feed acts of violent terror, and conversely, how violent terror can feed and perpetuate online propaganda." She likens social media platforms to "round-the-clock white supremacist rallies, amplifying and fulfilling their vitriolic fantasies."

Social media services also have fueled the noxious spread of hateful disinformation and conspiracy theories. Conspiracy theories are certainly nothing new. Observers have regarded antisemitism—which dates back millennia—as not merely the world's oldest hatred but as "the most durable and pliable of all conspiracy theories." *The Protocols of the Elders of Zion,* concocted in czarist Russia and dating from the early twentieth century, is a vicious piece of fiction that depicts an alleged Jewish plan to take over the world. Playing on long-standing antisemitic tropes, it has been used by Hitler, Stalin, Khomeini, Hamas, Hezbollah, and many others to justify the persecution and murder of Jews.

In America, conspiracy theorists have long targeted other groups besides Jews. During the early nineteenth century, rumors about nefarious Freemasons, Catholics, and even the so-called Boston Aristocracy swirled in our nation's political discourse. But conspiracy theories and misinformation are experiencing a new and disturbing heyday, spreading on social media, where they are rampant, and often inspiring demagogic political leaders who wield them as weapons in the real world.

The most obvious example is QAnon, which one expert on genocide has linked to Nazism and analyzed as "a rebranded version of the [*Protocols*] combined with the Blood Libel." (The latter is the classic antisemitic notion that Jews kill gentile children and harvest their blood for use in religious rituals.) But many other conspiracy theories make their way around the internet's nether regions, among them the contention that the 2020 presidential election was rigged, that Jews are manipulating the COVID-19 vaccine to achieve world domination, and that "'global elites' will use the pandemic to advance their interests and push forward a globalist plot to destroy American sovereignty and prosperity." This last theory, known as "the great reset" and popularized by hardened conservative commentators like Tucker Carlson, Laura Ingraham, Glenn Beck, and Alex Jones, is implicitly antisemitic (in its reference to globalism and elites, which is a classic and enduring trope for

Jews) and often nakedly so. Administrators of social media platforms know about these conspiracy theories; they promise to take the posts off, but a good chunk of the time, they inexplicably fail to act.

Earlier in my career, I experienced the sheer malice of conspiracy theories firsthand.

On April 3, 1996, Commerce Secretary Ron Brown and dozens of others perished when their plane crashed in Croatia. Secretary Brown was my boss in the Clinton administration. A number of those victims were my close friends and colleagues.

After the crash, right-wing media peddled conspiracy theories claiming that President Clinton had orchestrated the crash because Secretary Brown had damaging information about him. Those lies were not merely idiotic but utterly profane, a despicable desecration of the memory of these individuals. To this day, I remain outraged that people would stoop so low.

It's notable, though, that this poison didn't spread more rapidly and penetrate more deeply; that certainly would have been the case had social media been around in 1996. Though the mainstream media was far from perfect, it didn't prop up such nonsense. Serious journalists, fact-checkers, seasoned editors, responsible executives, even ombuds-men—all these roles combined with a culture of professionalism helped prevent such vile rumors from taking hold. Today the absence of such tripwires allows conspiracy theories to spread far and wide on social media platforms before anyone even thinks to pull the fire alarm.

Conspiracy theories go beyond morally reprehensible; they are deeply dangerous. They help radicalize individuals and inspire them to perpetrate acts of violence. At the broadest level, they also erode trust in society and in democracy, catapulting people into a fractured "post-fact" world in which individuals not only can't agree on policy issues but can barely find common ground on what's true and false.

As Anne Applebaum suggested in her book *Twilight of Democracy,* the dissolution of public discourse into a miasma of "false, partisan, and often deliberately misleading narratives," fueled in part by social media, is helping to create an opening for authoritarianism. Feeling anxious and disconcerted, some members of the population chafe against the complexity and disorientation all around them and seek "new political language that makes them feel safer and more secure." As the historian Timothy Snyder pithily remarked, "Post-truth is pre-fascism."

## Reluctant Action

Recently, after years of inaction, the platforms have seemingly started to address the threat posed by extremism. Beginning in 2018, Facebook and Twitter removed some extremist and white-supremacist content; Twitter permanently barred the conspiracy theorist Alex Jones from its service, and Facebook promised to ban posts containing white-nationalist and white-supremacist messages. Facebook also created the Facebook Oversight Board, an independent panel of experts recruited from around the world and charged with helping the company make its most difficult decisions about contested content.

But as many have noted, Facebook is a private business and doesn't need such a quasi-governmental body. It could simply decide to enforce its own standards more effectively and adhere to similar norms that you find in nearly all other businesses in all other industries. By creating such a "supreme court," Facebook seems to have kicked the can down the road and punted its toughest issues to other people.

We did see more meaningful action in late 2020 following Stop Hate for Profit, a highly effective corporate accountability campaign organized by ADL and other civil rights and advocacy groups. ADL had been dealing with Facebook administrators for years, engaging with them at times on a weekly basis to call out hateful content and request their assistance in pulling it down. And then in 2019, the renowned comedian Sacha Baron Cohen delivered keynote remarks at ADL's annual Never Is Now conference. In just over twenty minutes, he dismantled Mark Zuckerberg's preposterous claim that the company was treating issues of free speech responsibly. Cohen's impassioned and entertaining observations captured global attention and kindled a spark in me.

Several months later, I invited Cohen and several other activists committed to changing the social media status quo to my office. Investor Roger McNamee attended, as did ex-Googler Tristan Harris and Common Sense Media founder and CEO Jim Steyer. We talked through strategy and agreed that a corporate-accountability campaign, if executed well, could prove effective.

After George Floyd's brutal slaying, ADL analysts spotted white supremacists communicating and organizing openly on Facebook groups to disrupt the protests then taking place across the country. I shared this information with Jim as well as with friends at NAACP, Color of Change, Free Press, Sleeping Giants, and other groups. We all agreed to take action. We then launched Stop Hate for Profit with a full-page ad in the *LA Times* asking companies to step back from advertising on the platform for the month of July unless Facebook complied with a simple, straightforward set of demands presented on the Stop Hate for Profit website.

When we announced the campaign, we didn't have a single company lined up to join us, but within weeks more than twelve hundred businesses and organizations had signed on to our cause. Whether they announced their efforts and tweeted out our hashtag or simply followed our lead without referencing our campaign, iconic brands such as Coca-Cola, Disney, Levi's, McDonald's, Starbucks, Unilever, and many others agreed to hit the Pause button on their Facebook advertising for a month. It was a mass exodus the likes of which Facebook had not seen in its fifteen-year history.

Although, as anticipated, the protest didn't dent the company's profits, it yielded massive reputational damage. With Sacha's help, we executed an action on Instagram a few months later with the support of celebrities like Leonardo DiCaprio, Kim Kardashian, Jamie Foxx, Katy Perry, and many others. We gave state attorneys general guidance that informed their landmark antitrust case against Facebook. We briefed members of Congress ahead of several hearings during which prominent tech executives were grilled. In the end, our campaign generated billions of impressions and galvanized a global movement that Facebook simply couldn't ignore.

The pressure exerted by that movement yielded results. During the months following the launch of Stop Hate for Profit, Facebook began to implement a series of measures that previously had seemed unimaginable. It started to remove racist and hateful content more aggressively. It took down armed militia groups. After ten years of refusals, it began to classify Holocaust denialism as hate speech, granting itself the authority to remove such lies from the platform. Facebook also released a "civil rights audit" of the platform conducted by a former ACLU executive, hired a vice president for civil rights to bring an internal perspective on these issues into its leadership structure, and agreed to participate in a third-party audit of its hate content.

While Facebook denied that our campaign prompted any of these steps, that just isn't credible. Reports we received from Facebook insiders revealed the impact we'd made. A senior leader—the head of Facebook's global advertising business—said that the summer of 2020, when our campaign took place, was "very rough" and "the most difficult summer I've ever had professionally for sure."

But Stop Hate for Profit had an even wider impact. After we launched our campaign and as advertisers and critics savaged Facebook for their lack of action, other companies took action. Reddit shut down more than two thousand message boards. Twitter cracked down on links to hate speech and ejected former KKK leader David Duke. Snapchat, TikTok, Twitch, and YouTube also announced actions to curb hate speech on their services. These actions weren't nearly enough, but they were long overdue.

## Clamping Down . . . Sort Of

Efforts by Facebook and others to clamp down on extremism and hate speech took on more urgency following the January 6, 2021, insurrection. The evening of the attack, I reached out to our coalition partners, and a few days later we released a statement calling on Facebook, Twitter, and YouTube to ban Trump permanently from their services on account of his serial lies about the election's legitimacy and his calls to thwart the democratic process. The evening after our statement's release, Twitter announced it would ban Trump permanently from its platform. Facebook had suspended him "temporarily" in the hours after the attack but soon clarified that the time frame would be indefinite. YouTube followed suit a few days later, announcing a thirty-day ban that was extended and remains in place as of the writing of this book.

And then they and other social media companies did even more. Twitter took down tens of thousands of accounts supporting the QAnon conspiracy theory, while Facebook removed hundreds of sites "linked to armed movements." Firms that supported the value chains behind these websites also stepped forward. Apple and Google booted the Far Right social media service Parler from their app stores, and Amazon announced its intention to stop serving the company's website. Together, these two moves rendered the service practically undiscoverable and incapable of operating.

While we at the ADL welcome such actions, let's not give these com-

panies too much credit. Their years of dawdling and denying responsibility allowed hate to take root in the first place. Their algorithms
enabled this intolerance to expand and spread across the internet, en-
trancing users who often didn't even realize what they were encountering and feeding this content to those who affirmed their interest.

And even now, the industry's collective response remains spotty and
insufficient. As of this writing, Twitter continues to allow on its service world leaders like Iran's Ayatollah Ali Khamenei, who frequently
tweets out blatantly antisemitic messages, and others who call for Israel's destruction. This incitement to violence, which in effect encourages
the murder of millions of people, falls far short of what should be per-
missible regardless of a person's political stature.

YouTube, Facebook, and Instagram continue to host the rantings of
the racist, antisemitic conspiracy theorist and musician Young Pharaoh
(if you haven't heard of him, his lyrics include garbage like "The Jews
sacrificing babies, it's really real"). And despite Facebook's pledge to
remove Holocaust-denial content, it lagged in actually taking action,
earning it a D on a 2021 Holocaust-denial-management report card
that ADL issued to social media companies. Overall, our research has
found that levels of hate online in the spring of 2021 were essentially
unchanged from 2020 despite the promises made by the platforms.

One media expert observed that the social media giants are "creeping
along toward firmer action," but, as I know firsthand, their hearts are
still not in it. I've had numerous meetings over the years with Mark
Zuckerberg, Jack Dorsey, and other social media executives. I've often
heard promises to take action and a litany of excuses, but in the end, I've
seen very little meaningful action.

In July 2020, for example, as Stop Hate for Profit gained traction,
leaders of our coalition sat down on a Zoom call with Zuckerberg,
Sheryl Sandberg, and others from the company. It was the same old

story. They wouldn't commit to any new action to address the problem of online hate. At one point, Zuckerberg remarked with pride that Facebook's AI had improved to such a degree that the company removed 88 percent of the hate speech before users even saw it.

I was dumbstruck by his self-satisfaction. I responded by noting that I had once been an executive at Starbucks, and we could never have gotten away with proudly claiming that 88 percent of our drinks didn't contain poison. Hate speech is precisely that — poison. And removing 88 percent of it isn't nearly good enough. On the contrary, it's scandalous.

These companies are among the wealthiest ever to exist on planet Earth. They have achieved the kind of success that preceding generations of firms could never have imagined, building user bases larger than the populations of any country in the history of humanity. And they've done so in less than two decades, generating hundreds of billions of dollars in profits along the way. Considering the extraordinary technical challenges that they have overcome, these businesses indisputably have both the engineering and financial resources to neutralize the problem of hate and do it now. And yet, they seem almost inexplicably unwilling to attend to basic expectations of product safety and customer service that virtually every other business abides by.

I mean, think about it. In 1982, a number of people in Illinois died after consuming Tylenol laced with cyanide. The manufacturer, Johnson and Johnson, recalled thirty-one million bottles of its product across America, redesigned the packaging to be tamper-resistant, and then put it back on the market, spending more than one hundred million dollars all told.

By contrast, extremists have used Facebook groups for years to spread antisemitic and racist ideology. They used it to plan the January 6 insurrection and the violence that disrupted the Black Lives Matter protests. And Facebook hasn't bothered to fix it. For that matter, the industry at large hasn't bothered to take a number of basic and practicable steps to address hate, such as fixing algorithms that feed users ever more extreme content, giving victims adequate tools to report harassment, transparently reporting intolerance on their sites, and reliably enforcing their own policies banning hate speech.

Social media companies could make their platforms much safer for all of us if they wanted to. They could slow our society's rise up the Pyramid of Hate. They just don't seem ready to do that.

## Big Ideals, Big Dollars

You might wonder why these companies aren't ready. I think there are at least three big reasons worth mentioning:

### Reason #1: These platforms are allergic to policing speech

Silicon Valley traditionally has fetishized an especially extreme form of libertarianism, one that touts the virtues of an utterly unfettered marketplace of ideas and shudders at anything smacking of centralized control over individuals, including restrictions on speech. A famous expression of this ethic appears in "A Declaration of the Independence of Cyberspace," a tract written by the iconoclastic poet, activist, and one-time Grateful Dead lyricist John Perry Barlow; I remember first reading it in the pages of *Wired* magazine decades ago. Barlow decried any regulatory involvement in cyberspace, calling governments "weary giants of flesh and steel" and imploring them to "leave us alone." He went on to claim that the denizens of cyberspace were "creating a world where anyone, anywhere may express his or her beliefs, no matter how singular, without fear of being coerced into silence or conformity."

That sounds noble enough, but it's quite radical and even ridiculous to advocate a world in which "anyone, anywhere may express his or her beliefs." Anyone? Really? The Supreme Court has laid out very clear limitations to the First Amendment. You can't yell "Fire" in a crowded theater, nor can you legally incite a mob to lynch a Black person. As important as free speech is, it's not an absolute. Other social goods—such as everyone's basic right to life and limb—matter too.

Confronted with calls to remove hateful content, social media companies typically point to their principled belief in freedom of expression. If speech isn't expressly prohibited by the First Amendment, they don't want to remove it. That sounds noble, and we at the ADL have always been staunch supporters of the First Amendment. We recognize that a great deal of hate speech *is* constitutionally protected, which means social media platforms bear no legal obligation to remove it.

But how about a moral obligation? Media platforms aren't legally obligated to remove disgusting hate speech, but as private actors, they also aren't legally obligated to keep it on their platforms, let alone prioritize it as clickbait to attract users. If you're a white nationalist and you want to hold a convention, most hotels will turn you down, and that's their legal right. If you try to put up flyers promoting a racist book club in a company cafeteria, most businesses can and will bar you from doing so. Social media services can do this too, but because of their libertarian fantasies, they historically have abdicated any accountability, leaving society to pay an increasingly steep price.

As Steve Huffman from Reddit once remarked to me, "There has always been a fringe element in society. The trick is to keep it on the fringe." I think that sums up how companies could navigate this complex issue. If they insist on a dogmatic approach to free expression that exceeds constitutional requirements and even moral bounds, then social media platforms could allow the worst elements to post their poison but make it impossible to find. Take the content out of search. Don't promote hate through your algorithms. Remove all monetization opportunities. These and other simple technical fixes could allow social media platforms to have their proverbial cake and eat it too.

### Reason #2: There's no real economic punishment for hosting hate speech

During the 2020 election, right-leaning news outlets gave voice to unfounded rumors that voting systems in several states had been tampered with in order to help Joe Biden defeat Donald Trump. In March 2021, a maker of voting machines, Dominion Voting Systems, sued Fox News for $1.6 billion, alleging that "Fox sold a false story of election fraud in order to serve its own commercial purposes, severely injuring Dominion in the process." Another voting-system company, Smartmatic, filed a similar lawsuit, asking for $2.7 billion in damages.

Do you know why traditional media outlets like NBC News, CNN, and the *New York Times* don't feature members of extremist groups peddling false and hateful conspiracy theories or advocating for the removal of all Black, Latinx, Asian, or Jewish people from the country? Because if they do, they risk getting sued by people injured by this speech. In our society, you can lie about people, defame them, and demonize a group all you want. But be prepared to lawyer up when they come after you for damages and then pay up when the courts award them.

Social media companies don't have to worry about lawsuits from, say, victims of attacks perpetrated by extremists who were radicalized on their platforms. That's because these platforms are shielded by a specific provision, Section 230 of the Communications Decency Act, enacted by Congress in 1996. Section 230 states: "No provider or user of an interactive computer service shall be treated as the publisher or speaker of any information provided by another information content provider."

From newspapers to networks, billboard companies to radio opera-
tors, all forms of traditional publishers bear liability for harms done by
content disseminated under their editorial control. With some exceptions (such as child pornography), Section 230 allows internet companies to host user content and moderate that content without risk of being sued. Court rulings have specifically affirmed platforms' immunity
for speech by extremist groups. An example: a 2019 case upholding a
lower court's ruling that Facebook was not liable for violent attacks coordinated and encouraged by Facebook accounts linked to Hamas, the
militant Islamist group.

Proponents of the law laud it as an essential safeguard of freedom
of expression. Social media companies like it because they can make
money by posting content created by users and accompanying it with
advertising. If you're a consumer, you might like it too; Section 230
has facilitated the creation of crowdsourced services that we've come to
rely on, like Craigslist, Yelp, Google Maps, and Wikipedia. But at the
same time, consumers have no legal recourse when they encounter hate
speech on these services. Victims can't sue Facebook or Twitter to make
the platform take the speech down and compensate them for any harm
they sustained.

Without a real economic incentive, social media companies didn't
bother to build safeguards into their core products. Imagine a new car
delivered to customers without seat belts, airbags, or other basic safety
features. Imagine a new lawn mower without an outer metal casing
that separates the user from those deadly sharp, rapidly rotating blades.
That's what social media sites delivers—a product with negligible
safety features. They don't even maintain the kind of customer service
department that would allow you to seek support and talk to a live person when you're being harassed on their service, let alone when you encounter disturbing content. Have you ever heard of 1-800-Facebook?
I didn't think so.

Our society's legal regime has induced these companies to design their products to enable the free flow of filth. Without any real product safeguards, extremists have an easy time infiltrating social media sites and preying on people. It truly is the Wild West—John Perry Barlow's libertarian dream but, increasingly, our collective nightmare.

### Reason #3: Companies don't want to mess with their cash cow

But these platforms aren't simply akin to cars without seat belts. They're more like cars in which it's impossible to install seat belts without changing their core function. Deep in the bowels of Facebook and Twitter, powerful algorithms determine what content users see in their feeds. Since these and other social media services make money from attracting users and then selling those eyeballs to advertisers, they quite understandably designed their algorithms not for safety but for another end entirely: user stickiness.

The more time a user spends on these sites, the more revenue the services generate from advertising, and the more money they make.

As my friend Tristan Harris and numerous other observers have pointed out, these algorithms keep you glued to your screen by feeding you more of the content that they think you want. Human psychology being what it is, the algorithms feed you content that increasingly is more extreme, provocative, angry, and sensationalist, since that's what drives clicks and keeps users on the service. If you start out reading right-wing content that is mildly racist or antisemitic, algorithms likely will expose you to more of the same as well as content that is increasingly explicit and violent. In this way, they serve as engines of polarization and pathways to radicalization.

A former Google employee explains it this way: "There's a spectrum on YouTube between the calm section—the Walter Cronkite, Carl Sagan part—and Crazytown, where the extreme stuff is. If I'm YouTube and I want you to watch more, I'm always going to steer you toward Crazytown." A Facebook team acknowledged problems with the company's algorithms in an internal presentation, informing top leaders at the company, "Our algorithms exploit the human brain's attraction to divisiveness." The landmark documentary *The Social Dilemma* lays out this case with terrifying accuracy, sourcing insights from more than a dozen former social media executives whose credibility on these issues is unimpeachable.

As companies tweak their algorithms over time, they do so in ways designed to increase user engagement, with scant attention paid to the implications for hate speech. "YouTube [has] purposefully ignored warnings of its toxicity for years — even from its own employees — in its pursuit of one value: engagement," one scholar observed. The same could be said for other social media platforms.

Social media services that really want to make their products safe wouldn't just nibble around the edges. They would redesign the algorithms to steer people away from hate. But that likely would result in reduced user engagement to some extent and, with that, reduced revenues. Are publicly traded companies like Facebook and Twitter eager to do that? Not very. As one expert suggests, Facebook "always [does] just enough to be able to put the press release out. But with a few exceptions, I don't think it's actually translated into better policies. They're never really dealing with the fundamental problems."

To paraphrase President Biden's message to Facebook and other social media companies: Your product is killing people. Take it offline and fix it, even if it costs you. It's your social and moral obligation.

## Common Excuses Social Media Executives Make When Confronting Demands to Remove Hate Speech

1. "It's simply too hard to do at scale."

*Really? Facebook made profits of $32 billion in 2020. I think they've got the resources to figure it out.*

2. "We're not perfect, but we're on a journey."

*A journey, huh? See item number 1. You've got the resources. Stop making excuses.*

3. "We're doing our part."

*No, you're not. And most of America doesn't think you are either.*

4. "We don't want to inconvenience our users."

*What about people murdered by extremists radicalized on your services—do you care about inconveniencing them?*

## Why I'm So Emotional

You might think I'm being excessively shrill in my critique of social media. Shouldn't I just get off my soapbox, giggle at the videos of cute kittens, and relax a little? Is it really that bad?

After one meeting with the managers of a social media company, I heard from a friend who was in touch with an executive on the team; that person had been struck by my tone in the meeting and shared that, in this executive's opinion, I was "very emotional."

My response: guilty as charged.

Yes, I'm emotional, and for good reason: because hate is spreading, people are dying, and society is moving rapidly up the Pyramid of Hate.

If this executive had accused me to my face of reacting too vehemently, I would have said right back: Why aren't you *more* emotional? Indeed, why aren't *all* these executives more mindful of the damaging social impact of their products, and why aren't they inspired to do something about it?

I'm not at all alone in my concerns. Our surveys show that the vast majority of Americans are concerned about hate online and want to see social media take meaningful action. Following the January 6 insurrection, 61 percent of respondents in an ADL survey felt that social media platforms were "significantly" or "somewhat" responsible.

If you think I'm overly emotional in worrying about the danger of social media, consider this: It seems clear that Facebook has already had a hand in a recent genocide. In Myanmar, military forces and mobs in the majority Buddhist country killed thousands of Rohingya Muslims and forced hundreds of thousands of others from their homes, acting with "genocidal intent," according to a United Nations report. UN experts state that Facebook, which is widely used in the country, had hosted pages associated with the military that had spread hate toward the Rohingya and incited attacks "for months, if not years, before Facebook finally took action against them."

As one expert claimed, Facebook has "substantively contributed to the level of acrimony and dissension and conflict, if you will, within the public. Hate speech is certainly of course a part of that. As far as the Myanmar situation is concerned, social media is Facebook, and Facebook is social media." After receiving the results of an independent report the company had commissioned on its presence in Myanmar, Facebook itself admitted that it had to "do more to ensure we are a force for good in Myanmar, and in other countries facing their own crises."

Social media platforms in general must do more to prevent extremists from hijacking their online communities and inciting violence. We've already seen the role that social media played in laying the groundwork for the violent January 6 insurrection. What more needs to happen in the United States and elsewhere before the industry truly reforms itself? Do we need to see widespread civil unrest, an explosion of mass-casualty events, the rise of an authoritarian dictatorship, the creation of concentration camps to house hated minorities, or organized genocidal acts?

Yes, I'm emotional. After reading this chapter, I hope you'll be too. And if you're not, just consider what the future could look like if hate continues to explode.

# 6

# AMERICAN BERSERK

In 1835, Francis McIntosh, a steamboat cook and a free Black, was arrested in St. Louis, Missouri, accused of preventing two law enforcement officers from doing their jobs. A justice of the peace ordered those officers to escort McIntosh to jail. Upon learning that he could face five years in prison, McIntosh tore away from the officers and attacked them with a knife, killing one.

Locals captured McIntosh and brought him to jail. Word got out of what had happened, and outrage spread among the population. "The excitement was intense," a newspaper report from the time related, "and soon might be heard above the tumult the voices of a few exhorting the multitude to take summary vengeance."

Take vengeance they did. A mob broke McIntosh out of jail, chained him to a large tree, surrounded him with wood, and set him on fire. It took twenty minutes for him to burn to death. Afterward, a band of youths threw stones at his corpse, competing to hit and break his skull. By some accounts, it was the first lynching in American history.

The death of Francis McIntosh might have fallen into obscurity had it not been for a young lawyer named Abraham Lincoln. In his famous speech "The Perpetuation of Our Political Institutions: Address Before the Young Men's Lyceum of Springfield, Illinois," Lincoln described

McIntosh's death and other instances of mob violence as an existential threat to American democracy. As he pointed out, America enjoyed ample blessings, including fertile soil and an enviable system of governance bestowed by an earlier generation of revolutionaries. In scanning the horizon for potential threats, Lincoln dismissed the idea that a foreign power could stamp out American democracy. But there was another danger worth ruminating over: Americans themselves. A mortal threat to the nation "cannot come from abroad. If destruction be our lot, we must ourselves be its author and finisher. As a nation of freemen, we must live through all time, or die by suicide."

Lincoln was fairly specific about the form such "suicide" might take. It was disregard for the rule of law, "the growing disposition to substitute the wild and furious passions, in lieu of the sober judgment of the Courts." Left unchecked, mob rule could prompt citizens to lose faith in their government, leaving the country vulnerable to opportunistic would-be authoritarians.

Donald Trump's presidency and the January 6 insurrection make Lincoln's words seem especially prescient. But while Lincoln focused on the threat posed by abandonment of the rule of law, others have observed a related danger: a peculiar undercurrent of hate visible in American society. I think in particular of the late novelist Philip Roth and his notion of the "indigenous American berserk." As one cultural critic summarized it, the phrase "suggests that American violence is not only native to Americans but intrinsic to them; that it lurks just beneath the surface and can be set off at any time; that it partakes of the madness particular to mobs; and that you ignore it at your peril."

The ADL's data doesn't lie; the American berserk has experienced a resurgence of late. But where might it take us? Will we see religious minorities marginalized and attacked in a spate of one-off attacks? Will we see more mass shootings targeting racial minorities? Will we see undocumented minorities interned in detention centers on our borders? Or will it be worse—sectarian violence, full-blown civil unrest, mass violence, even a genocide killing thousands or millions?

It might sound hysterical for me to even pose such questions. But my Jewish identity and my role leading a Jewish anti-hate organization compel me to do so. By the time you finish this chapter, I hope you'll agree that such scenarios are possible and must be considered. If we're to understand that threat and mobilize to stop it, we must understand to the fullest what is truly at stake.

## Take Nothing for Granted

I attended a dinner in the spring of 2021 and happened to sit beside a man whose intersecting identities made him a target for animosity many times over. A native Canadian, he was an Ismaili Muslim (Ismailis are a small sect of Shia Islam with a global population about as large as that of the Jewish community). During our conversation, he also identified himself as South Asian, a member of the LGBTQ community, and married to a Jewish man. As we ate, we spoke about the contemporary situation in America. He told me he was optimistic about the position of minorities in the United States now that President Biden had taken office. Although there might be some bumps, he felt that our civil society would hold together, normalcy would reassert itself, and everything would be fine.

I told him I wished I could agree, but Jewish history suggested otherwise. Jews have lived as minorities in numerous societies throughout the ages, and with relatively few exceptions, they've eventually confronted a dire fate. In nearly every context, their honeymoon in their newfound homelands came to a precipitous end—they were exiled, compelled to convert, or slaughtered. For this reason, many Jews today find it fairly easy to envision nightmarish scenarios. The historical trauma of fleeing every four or five generations has been encoded into our communal DNA, leading to a penchant for catastrophic thinking.

Sometimes people mock this Jewish tendency—we've all heard jokes about the anxiety of Jewish mothers—but its roots are very real. Jews might feel deeply patriotic and enjoy the peace and security of living in a free society like America, but by no means can we take it for granted.

As I reminded my dinner companion, it was indeed reassuring to see Donald Trump leave office, but he had very nearly gotten reelected. This president severely mishandled the COVID pandemic, leading to the deaths of hundreds of thousands and severe economic dislocation. Despite that, over seventy-four million people—47 percent of the electorate—had voted for him. If just a few more counties in the United States had gone Trump's way, he would have won.

I'd add that fringe elements on both sides of the political spectrum, not just those on the right, now pose serious threats. Many have commented on the galloping illiberalism in some quarters of the political left—cancel culture stifling debate, journalists reimagining themselves as activists and pursuing agendas rather than reporting facts. Such critiques sound abstract, but they touch on something real.

During the spring of 2021, when fighting erupted between Israel and the Hamas government in Gaza, numerous elected officials and community leaders on the political left attacked Israel by spreading outright fictions. They contended, quite outrageously, that the Israeli military was intentionally murdering children, that the Jewish state was committing genocide along the lines of Nazi Germany. In a particularly offensive case, a Democratic political candidate in New York wrote comments about "fake Jew Zionists" in a post notably still up on Facebook at the time of this writing. She opined that "Zionists are a very select group, with tyranical [sic] political aims disguised as a religious/geographic/ethnic group." She continued by employing a typical extreme Far Right talking point about ZOG, or the "Zionist occupied government," a term used to connote nefarious Jewish control over the highest levels of politics: "We are all under attack from the ZOG. Both parties work for the zionist usury banksters, the small group of tyrants who run the FEDERAL RESERVE and have usurped our de jure (right and just) beloved country."

And we have seen other inflammatory gestures of anti-Israel advocacy that, while not necessarily antisemitic, gnaw at the norms of civil discourse. Take, for example, the creeping mainstreaming of the destruction and burning of Israeli flags, an instance of which occurred in midtown Manhattan during the writing of this book. On college campuses, we've seen multiple instances of anti-Israel students removing or otherwise vandalizing American flags in just the last couple years. I want to stress here that such action on a Palestinian flag or any other national symbol would be equally condemnable. But this is happening in an environment of heightened antisemitic harassment and violence in which Jews feel increasingly vulnerable.

At the risk of stating the obvious, I must repeat that incessant inflammatory rhetoric against the Jewish state can imperil all Jewish people. The Israeli government is not exempt from criticism, but such incendiary and false claims unsurprisingly incite tensions and spark outrage. They can quickly escalate — as in the spring of 2021, when they touched off a wave of harassment, vandalism, and violence targeting Jews in America and around the world. As noted earlier, ADL tracked a week-over-week surge of antisemitism in America in May 2021, with incidents spiking 115 percent year over year. The numbers were even worse abroad, particularly in Europe. The United Kingdom, for instance, saw a more than 500 percent rise in anti-Jewish attacks.

Conspiratorial claims about a country or community can prompt hate and violence against any group. Indeed, consider the 2020 surge in attacks on Asian-Americans that took place as political leaders recklessly blamed China for the advent of COVID-19. Likewise, the murderous August 2019 attack in El Paso, Texas, directed against people perceived to be Mexican was a heinous xenophobic crime inspired by elected leaders' and political pundits' rage against Latino immigrants.

I'm not a futurist, so I can't talk about pending events with anything like certainty. Perhaps American life will tick along just fine, and a decade from now we'll look back on our current period as an unfortunate but not terribly consequential blip. My fear is that if we allow hate to go unchecked, we might well look back on this moment as the turning point at which America became unraveled and the American berserk came into ascendence.

I believe the worst scenarios — full-scale genocide, a conflict that rivals the Civil War in its bloodiness, the permanent destruction of American democracy — are highly unlikely. And I'm by no means alone. The United States Holocaust Memorial Museum's Early Warning Project assesses the risk in specific countries that mass killings of one thousand or more civilians will occur over a twelve-month period. According to the project's methodology, the United States had a 1 percent chance of seeing a mass killing begin in 2020 or 2021 and ranked sixty-sixth out of 162 countries in terms of risk.

That sounds pretty good, but before you entirely brush off talk of a cataclysm, there are other facts to consider. As low as it is, the risk of a mass killing is not zero. In fact, almost one hundred countries have a *lower* risk of a mass killing in the next twelve months than the United States. Other scholarship sheds light on the factors that might incline the U.S. toward a disaster.

Genocide Watch, an organization founded by Gregory Stanton, now a retired professor of Genocide Studies and Prevention at George Mason University, tracks genocide around the world, conceiving a society's movement toward genocide as a ten-stage process that includes constituent processes such as the classification of certain groups as Others; the naming of these groups and designation of symbols attached to them; the pursuit of discrimination against targeted groups; the dehumanization of others by designating them "aliens," "criminals," or "terrorists"; the organization of armed forces that could carry out a genocide and/or takeover of society; the polarization of society; the planning of a coup d'état; the persecution of targeted groups; the extermination of members of these groups; and, throughout the other steps and after them, the denial of wrongdoing.

As of this writing, Genocide Watch finds evidence for six of the ten stages in the United States, citing specific features such as the dehumanization of immigrants as rapists and robbers, the organization of white-supremacist groups, the polarization of Americans through racist and militarist propaganda, the spread of neo-Nazi conspiracy theories like QAnon, the development of a "leader" cult around Donald Trump, and the violent invasion of the U.S. Capitol to prevent certification of the 2020 election.

Professor Stanton notes distinct parallels between present-day America and Nazi Germany. "I see white-supremacist groups in the United States as very similar to the Nazis," he says. "And what really scares me is the number of Republican politicians who are in denial about this danger and in fact even support it. Now that also is similar to the way the Nazi Party rose in Germany." Stanton further likens the big lie of Trump's election victory to Hitler's claim that Jews and other civilians betrayed Germany during World War I, causing it to lose the war.

Imagining how a conflagration might unfold, Stanton describes a scenario in which a crisis—a natural disaster (such as a pandemic or weather-related event) or civil unrest (narco-terrorist takeover or civil war)—drives millions of desperate migrants to flee to the United States' southern border. Despite the efforts of border control to keep them out, migrants cross the Rio Grande and the Sonoran Desert and enter the country. This prompts white supremacists and white nationalists to claim that immigrants are taking over and that whites must rise up and defend themselves. These extremists get a hearing from large numbers of right-leaning voters in the country. Thanks to restrictive voting laws, whites occupy most positions in power at the state and national level.

White-power advocates pass laws abolishing gun control and militarizing the police and American society. They establish detention centers for the unwanted immigrants and imprison the violent demonstrators who protest. When the camp prisoners try to break out, horrible atrocities ensue. "Never forget that Japanese-Americans were interned in camps during World War Two and their imprisonment was upheld by the U.S. Supreme Court. Canada did the same to Japanese-Canadians and Italian-Canadians. I don't think it's at all impossible in this country," Stanton says.

And neither do I, particularly after a 2018 visit I made to the U.S.-Mexico border, where I saw the detention centers that the U.S. government had set up to house migrants intercepted there. Despite the overheated rhetoric of some members of Congress, these certainly were nothing like Nazi-era concentration camps. And yet the misery and depressing conditions were undeniable. Persecution always starts somewhere.

## No Longer Unthinkable

My research team and I consulted several experts about the risks facing the United States. No expert we spoke with argued that genocide, a hate-fueled civil war, or some other decisive breakdown of American society was imminent or even likely. But they did point to some disturbing dangers on the horizon.

When I spoke with Yale professor Tim Snyder, he likened the recent spate of voter-suppression laws spreading across the U.S. to the Nuremberg Laws in Germany, which created two tiers of citizens—Jews and non-Jews—as a prelude to relieving Jews of their citizenship entirely. He worried that our contemporary laws could lead to racial stratification in America, resulting in large numbers of disenfranchised voters who overwhelmingly hail from communities of color.

Harvard political scientist Steven Levitsky articulated similar anxieties. In Republican-dominated states, he noted, state legislatures in the future might overturn local election results and enforce minority rule on an ethno-religious basis. Could that lead to openly discriminatory policies that infringe on minority rights? Could it lead to political violence? Professor Levitsky refused to speculate. But he did make two observations that give me pause.

First, he noted that we can't assume that conventional limitations on extremism will hold. "We've entered a world in the last decade that most of us growing up couldn't have imagined," he said. "I would have thought it unthinkable that we could create second-class citizens and legally empower white Christians. But the ethno-nationalist turn in the liberal West has been quite striking, and it could happen here."

Even more disconcerting, Professor Levitsky remarked that he didn't know of a "democracy on Earth that has made a transition in which a dominant founding or dominant ethnic group loses its majority or dominant status and democracy survived." But that is precisely what appears to be happening in the United States, with the eclipse of the traditional white Christian majority. The result could be destructive. "We seem to be entering a new period of violence," he said, "one not unlike other periods of crisis, whether it's Europe in the 1920s and early '30s or South America in the 1960s and '70s or the United States in the 1840s and '50s."

Will large-scale violence break out? Unlikely, given how entrenched the rule of law is and how powerful the state is. But it's not impossible either. Professor Snyder shared a similar sentiment, expressing his concern that this recent spate of voter-suppression laws, so clearly designed to disenfranchise voters of color, could trigger an ugly cycle of racial violence.

## Organized Chaos

Other experts worry that the risk of civil war in the United States, although still small, is growing. Barbara Walter, professor of political science at University of California, San Diego, and an expert on civil wars, observes that countries become more prone to civil wars when their political systems enter a middle zone between democracy and autocracy, combining elements of the two. Guess where the United States has been heading in recent years? Right into that middle zone, with the weakening of its democratic institutions. In her estimation, the country still has some ways to go before the risk becomes high, but it could very well happen.

Mention of civil war might conjure images of two armies going toe to toe at Gettysburg. Walter argues that any civil war that does break out in the United States would probably be more like what the world saw in Northern Ireland during the 1970s and 1980s, when members of the Catholic minority mounted an insurgency against the British government and various loyalist groups in turn attacked Catholics. That is, it would be a smaller conflict, more decentralized, fought by multiple competing factions, and with actors resorting to acts of terror. "You're going to have a bombing here, a sniper attack there—it's almost going to seem like organized chaos. We could have decades of sustained violence where you have a subset of the population using unconventional forms of violence to try to get what they want." Because the violence would be dispersed and clandestine and have popular support in some areas, the government would have trouble suppressing it entirely.

We potentially could see local white-supremacist militias in various regions of the United States target Jews, Blacks, and other minorities in an attempt to push them out and create mini-white-majority eth-no-states. That scenario, although it might seem unlikely, is not im-plausible when you consider the sharpening rhetoric, the proliferation of firearms, and the precedents of recent terror attacks in the homeland. At the very least, Walter suggests, we can expect in the years ahead to see more support for extremists by an embattled white population. "There's a lot of anger out there. That anger isn't going away. And it's now being fed this big lie of stolen elections."

I might observe a similar echo in the clashes that erupted in May 2021 after the Gaza conflict. There, we saw anti-Israel activists engaged in acts of intimidation and actual violence against Jewish communities in the United States and Europe. Those acts were sudden and seemingly uncoordinated, crimes of opportunity. Prominent elected officials re-peated statements against Israel reminiscent of the slanders regularly proffered by disgraced British politician Jeremy Corbyn — they were inflammatory and entirely one-sided in assigning blame. If such rhet-oric against the Jewish state continues to swirl and animus amplifies, it seems conceivable that this anger could coalesce into more coordinated violence against Jewish communities on par with the danger posed by the Far Right.

However unlikely the prospect of violent scenarios might be, Walter is worried. She shared with me that the specter of such potential acts led her to contemplate a plan B for her family — relocation to a foreign country. When an expert in civil wars starts to consider a plan B, I think that should make the rest of us worried too.

## A "Plan B" Isn't Enough

Clearly, we need to redouble our efforts to push back against the hate that threatens to destabilize our society. But how? Now that we appreciate the dire and growing threat that hate poses, let's turn to some important solutions.

You might think that the government should bear the bulk of the responsibility. Let's empower the FBI to infiltrate extremist groups before they strike. Let's demand that the courts hold social media responsible for spreading hate. Let's mobilize local law enforcement and the judiciary to ensure that perpetrators of hate crimes are apprehended and punished. We can and should take all of these steps. But despite its perceived power and authority, the federal government cannot solve this problem on its own. We need a "whole of society" strategy, activating a wide array of actors at all levels of government as well as in the private sector to push back against bigotry.

Most of all, we must mobilize as individuals to counter hate when we see it, model values of tolerance and inclusion, and serve as allies for those victimized by prejudice. Lincoln might have been right that the greatest threat to American democracy—and, by extension, to marginalized groups in America—is Americans themselves. But I would argue that we're also the ultimate protection against that threat. Each of us stands to lose if the haters prevail. But each of us has the ability to do what's right and make a difference in the fight for a safer, more civil, more humane society.

Yes, the American berserk has become more visible in recent years, but let's not allow it to destroy us. Let's push it back into the shadows where it belongs. Let's seize the moment, overcome our fears and doubts, and dedicate ourselves to bringing about a post-hate society animated by compassion, courage, and decency.

*Part II*

# DISMANTLING THE PYRAMID

# 7

# FIGHTING HATE IN EVERYDAY LIFE

There's a hateful story that white supremacists in America and elsewhere like to tell one another, and it runs like this: The white population is under siege. All those brown and Black people immigrating to our country are coming to take what's ours. Our jobs. Our homes. Our opportunities. The result, if left unchecked, will be nothing less than the destruction of the "white race"—a "white genocide." Oh, and you know who's behind it all? Those scheming Jews, of course.

If this crackpot story sounds vaguely familiar, maybe it's because you saw the torch-bearing mob march in Charlottesville in August 2017 shouting "You will not replace us" and "Jews will not replace us." Or maybe it's because you happened to notice that the perpetrators of the Tree of Life, Christchurch, Poway, and El Paso atrocities all posted manifestos after their attacks referencing versions of this story, known as the "great replacement theory."

Or maybe it's because you tuned in to Tucker Carlson's April 8, 2021, show on Fox News. Carlson has a history of publicly espousing xenophobic and anti-immigrant sentiments, often seemingly borrowed from white-supremacist message boards. That evening, while talking about immigration, he said,

> I know that the left and all the gatekeepers on Twitter become
> literally hysterical if you use the term "replacement," if you

suggest that the Democratic Party is trying to replace the current electorate, the voters now casting ballots, with new people,
more obedient voters, from the third world. But, they become
hysterical because that's what's happening, actually. Let's just say
it: That's true. Every time they import a new voter, I become
disenfranchised as a current voter.

When a member of my staff showed me a clip of Carlson's rant, I was
disgusted and terrified for our country. Here was a media personality
with 3.4 million nightly viewers and the highest-rated primetime cable
news program spewing a vile conspiracy theory that had already helped
spawn massacres of innocent people.

No, Carlson wasn't holding a torch and marching with a crowd of
angry extremists. What he did was worse. He was performing what I
call "hate laundering." By giving extremist theory a patina of respectability and reasonableness, he popularized it for the masses. A fringe extremist belief was moving from the margins into the mainstream before
our very eyes, raising the odds that America would proceed still farther
up the Pyramid of Hate.

All of us as individuals have a duty to stand up and make ourselves
heard when we encounter hateful discourse. In this instance, I did exactly that. The day after Carlson's tirade, I wrote an open letter to
Fox News CEO Suzanne Scott condemning Carlson for repeating
"a classic white supremacist trope that undergirds the modern white
supremacist movement in America." Given Carlson's long track record of making divisive statements, I implored Scott to take him off
the air.

This was not a request I made lightly. As the leader of a hundred-
year-old civil rights organization, I have the utmost reverence for the
First Amendment and believe that everyone is entitled to air his or her
opinion. And as I've explained elsewhere, I don't at all think we should

go around "canceling" one another every time we encounter ideas we don't like or agree with.

But just because Tucker Carlson thinks hordes of immigrants are replacing whites doesn't mean that Fox News, its owners, and its advertisers should give him a bully pulpit. If they care about viewers' safety and the state of our society, they shouldn't. And although it's important to give people a chance to repent and mend their ways, Carlson has been a repeat offender many times over. He's had ample opportunity to reform his thinking and yet stubbornly refuses to do so. Far from offering innocuous "comments on immigration," as some claim, he was peddling the kind of poison that has led directly to violence in Charlottesville, Pittsburgh, El Paso, and elsewhere. In unfortunate situations like this, taking a stand and demanding accountability is not only legitimate; it's necessary.

In response to my letter, Lachlan Murdoch, CEO of Fox News Corporation and the billionaire son of media mogul Rupert Murdoch, wrote back defending Carlson and refusing to take him off the air. Murdoch claimed that later in the segment, Tucker had actually repudiated replacement theory. That defense was pathetic—a point I made in a letter responding to Murdoch.

Realizing that I wouldn't get Fox and its owners to do the right thing on their own, I called on brands to exert pressure on Fox by pausing or pulling their advertising. "Commit to this fight," I told them. "Choose a side." As of this writing, the controversy is ongoing, the ultimate outcome unclear. We might not succeed in getting Carlson off the air, and if we do, there will almost certainly be other Carlsons to take his place. But again, we all have to stand up against hate. And in this chapter, I'll show you how.

## A Framework for Fighting Back

In calling out Tucker Carlson, I was deploying a simple but powerful framework ADL developed to help individuals find their voices and respond to hate. It's called "Speak Up, Share Facts, Show Strength."

*Speak up* means when you see something, you say something, even if it's uncomfortable. Muster the courage to step up and step out, even if you're alone.

*Share facts* means grounding your response in evidence and data. If you're engaging online, speak as calmly and respectfully as you would if the conversation were occurring face to face.

*Show strength* means digging deep and boldly defending yourself but also looking out for those in need of protection. Stand up for yourself and serve as an ally in situations when hatred doesn't affect you directly.

We designed the three Ss to encapsulate the ADL's one hundred years of experience fighting antisemitism and hate of all types. It's an easy-to-remember construct that helps our hundreds of staffers and thousands of volunteers across the country respond quickly and effectively when people ask them how to combat hate. No matter your age, profession, or stature in the community, you can use the three Ss to stop the virus of hate from spreading.

Let's take a closer look at what each step entails.


### Speak Up

It's truly important to interrupt intolerance at the moment when you see it happen and to do it intentionally and publicly. That could mean speaking up at the figurative or metaphorical watercooler at work when someone makes an inappropriate joke. At school, it could mean reporting hateful actions to a teacher or parent. On social media, it could mean flagging someone's tweet or post. At home, it could mean respectfully pushing back on family members who use hateful slurs.

Now, I'm not suggesting that you should *always* react on hearing or encountering hate. Not everything is a five-alarm fire, and you might find yourself in a situation where your personal safety is at risk. It's important to use judgment. However, most of the time, the human tendency is to avoid conflict, retreat from tension, and refrain from speaking up. Perhaps you fear being ostracized from a group as a result of pointing out something uncomfortable. Perhaps you fear causing offense and damaging your relationship with a perpetrator of bias or hate.

Whatever the case, it's important to work through these fears and concerns so you can confront issues in an appropriate way. When people around you are dismissing or ignoring behavior that is clearly beyond the pale and depraved, you owe it to yourself and your community to speak up. That might mean taking at least some risk. Maybe your tweets won't get quite so many likes. Maybe that colleague in your office will think you're a little prudish. Maybe you'll temporarily anger that family member or friend. But these small risks are worth taking if it makes your community safer and more humane. Indeed, one could see them as the reasonable price of being principled.

Let's say you encounter hateful speech in the course of having a conversation with someone. What should you say? Here are four strategies to try:

(1) *Ask a question:* If someone makes a hateful remark about a group of people, respond by saying, "What do you mean?" or "Are you referring to everyone who is [insert identifier], or are you speaking of someone in particular?" or "What are you basing that on?" Asking for clarification can prompt speakers to stop and ponder

what they're saying without you having to point it out directly. It's a soft way of stopping the conversation and drawing attention to hate rather than letting it go and become normalized.

(2) *Explain the impact:* Often people who utter a hateful slur or joke don't intend to cause grievous harm. They are simply repeating something they've heard without giving it a lot of thought, or they assume that slurs or comments are "only words" and not that meaningful. You can help by explaining the impact to them, saying something like "When you say that, it is really damaging to an entire group of people" or "Statements like that reinforce stereotypes that really harm people."

(3) *Invoke a broader context:* If someone makes a derogatory statement about a group, challenge the accuracy of the statement by prompting that person to see the world more expansively. Say something like "I don't think that's a [insert identifier] thing. I think lots of different people have that quality" or "You can't make a generalization about any one group of people based on your interaction with one person" or "Every human deserves respect and decency, don't you think?"

(4) *Connect to history:* Challenge hateful ideas by pointing out historical contexts of which they might be ignorant. Say: "What you're saying actually feeds into a really old trope of . . ." or "That language supports a legacy of disrespect and violence . . ." or "Let me explain how that language was historically used to talk about people."

It's one thing to call out hate speech during a conversation, but what do you do if you encounter someone actively bullying or harassing another person at school or work? Should you intervene or stay silent? Here are some questions to consider:

(1) *Is immediate action warranted?* Sometimes bystander intervention can cause even more pain or embarrassment to the victim of bullying. And you might be emotional yourself and thus less able to engage productively with the aggressor. Before rushing in, consider whether to take some time to calm down. If you can, talk with the targets to gauge their preferences. If you wait to take action, make sure that the targeted individuals know that you support them, and tell them what you intend to do.

(2) *Am I putting myself in danger?* If intervening will put you in harm's way, don't do it. If you're a student at school and feel uncomfortable or unsure, tell an adult. Always consider what the impact of your intervention might be on the targeted person.

(3) *Should I ask for help?* If a targeted person is in immediate danger or if intervention on your part could escalate the situation, you might want to call the police. If you're a student in a school setting, consider going to a teacher, guidance counselor, or other authority figure.

(4) *Does the targeted person need help?* If you can, talk to the victims. What support do they feel they need to feel physically and emotionally safe?

If you do decide to intervene, first ask the aggressors to stop the behavior immediately. Ask them questions that prompt them to reflect on what they're doing (for example, "What did you mean by what you just said?," "I don't understand; why would you say that?," or "That was really mean. Why did you say that?"). Speak up about your own feelings about the impact of the bullying ("I'd appreciate it if you didn't say that word around me because I think it's offensive").

If you're in a workplace setting, ask your bosses or supervisors to intervene and tell the aggressors to stop and apologize for their conduct.

Also, ask them to communicate policies about bullying and harassment and to enforce the rules. Express to them how important it is to use this episode as a teaching opportunity in which everyone can come to understand the harmful impact of intimidation. If you're a student in a school setting, ask an authority figure to take these actions.

If you encounter hate in public places (for instance, antisemitic graffiti or a swastika or someone shouting slurs), report it to the police. Document the incident by taking a picture or a video and ask police to complete an incident report. Don't remove offensive material until after police have investigated. Even if the police eventually determine that no law was broken, preserving the offensive material will help to ensure that some sort of follow-up action will take place.

Go online and complete an ADL incident report. Our experienced staff can provide you with further support and advice tailored to the situation. Report graffiti or offensive flyers to a representative of the building or public space where you found them. If possible, turn the incident into a teachable moment. Organize a small gathering, event, or conversation on the topic of mutual respect and understanding for all religious, racial, ethnic, and social-identity groups. A collective message that pulls together the community against a manifestation of hate can be very powerful and actually create a sense of cohesion and purpose.

If you encounter hate on social media, don't hesitate to call it out. Take screenshots of any harassment you receive. Call 911 if you feel you or someone else is in imminent danger. File a report on the FBI Internet Crime Complaint Center (https://www.ic3.gov), including as much information as you can. And be sure to report hateful content to the online platform (Facebook, YouTube, and so on) on which it appears. Once you've reported it to the platform, contact the ADL and schools or employers. If you're being harassed, resist the impulse to engage in conversation with the harasser. These interactions typically only ramp tensions up further.

When you encounter hate in traditional media, a good old-fashioned letter to the editor might be in order. Make sure the offending material really is hateful and not simply a viewpoint with which you disagree. Your local ADL office can help you make the distinction and can also help you determine if news stories are factually incorrect or otherwise problematic. When drafting a letter, do it quickly. The more time that passes, the less likely it is that the website or newspaper will publish it. Also consider leaving a message in the comments sections that many online news sites provide. Keep your letter brief and to the point, make sure it conforms to any guidelines the publication has, and avoid the trap of attacking the author of a news story personally.

If you see an ongoing pattern of hateful discourse, you might consider drafting a letter to the ombudsman or readers' advocate that many news sites employ. While social media platforms almost always lack ombudsmen, you can use their own services to call attention to a disturbing fact pattern. If you've tried unsuccessfully through formal channels to get the platform to take action, a few damning examples posted or tweeted for the public to see can often prompt a meaningful response.

### Share Facts

When you encounter hate, you might find yourself angry, even outraged. It's understandable to feel that way, but projecting those emotions can intensify intolerance rather than dissipate it. When responding to hate, it's important to do so in a calm, rational way that decreases tension. Wherever possible, take emotion out of it by sharing *facts* with others. Many biases, after all, are rooted in our ignorance of people who are different than us or in a misunderstanding of a culture or religion. By sharing facts, we lower ignorance and encourage reflection and understanding. We also establish our own credibility as knowledgeable interlocutors on the subject of hate.

ADL endeavors to be a fact-based organization. Go to our website, and you'll find numerous reports and blogs filled with the results of our original research and data drawn from other, publicly available sources. Every year, we release evidence-based reports, such as our audit of antisemitic incidents and our murder and extremism report. When I speak publicly, I make a point of referencing these reports whenever I can, grounding my positions as much as possible in empirical research rather than personal opinion. When you encounter hate at the watercooler or around the dinner table, you often won't have the facts you need at your fingertips. In these situations, responding with an anecdote that humanizes the issue can help.

If someone makes an offensive statement about the LGBTQ community, for instance, you might respond by noting that one of your good friends identifies as gay and would feel hurt to hear this said about him. If someone comments that Zionists are Nazis, you might respond by saying you have an Israeli friend whose parents are Holocaust survivors and that you can imagine this comment would be very painful for her to hear. In both situations, rather than laying out statistical data or historical details, you can enhance your response by grounding it in the empirical facts you *do* know about — those relating to the loved ones and respected colleagues in your orbit.

The important point here is to keep the conversation as dispassionate as possible wherever it is taking place. You need not forward that unhinged e-mail from your uncle's neighbor claiming some wild conspiracy that is undermining God-fearing America. You need not respond to that toxic tweet or angry WhatsApp message with something even more fiery. Try to be measured, not the loudest person in the room. It might feel emotionally satisfying to rant and rave and put offenders in their place, but if your goal is to convince people to disavow a prejudiced idea or even just question their underlying assumptions, you're doing more harm than good. When people feel attacked, they tend to react defensively, shutting themselves off and perhaps even lashing out further. You can stop this dynamic from playing out by staying calm and measured.

To become better at sharing facts, we all have a responsibility to educate ourselves about the various categories of identity and the persistence of certain prejudices. When it comes to Jews, for instance, do you know about the denominations that exist within the Jewish faith? Do you have a basic understanding of Jewish beliefs and practices? Do you know what antisemitism is and the many ways in which it has historically manifested? Are you familiar with the symbols and slurs that right-wing extremists use to delegitimize the Jewish people and that left-wing radicals use to delegitimize the Jewish state and anyone who identifies with Zionism? Do you appreciate how both sets of slurs often draw on hateful tropes that have lingered over centuries? Do you understand the arguments antisemites often make — for instance, the notion that Jews aren't patriotic citizens of the United States because their true loyalty is to Israel, or the idea that the Israeli lobby controls lawmakers in Washington, DC? Do you understand the difference between vicious anti-Zionist screeds about global control of political affairs and valid critiques of Israel, and do you know how to distinguish between the two?

Contemporary identities and the hatreds that arise around them are extraordinarily complex and always shifting. We constantly must deepen, broaden, and refresh our knowledge about Jews, Muslims, Blacks, Asian-Americans, Latinos, immigrants, LGBTQ people, and other targets of bigotry. When you reduce your own levels of ignorance, you're better able to spot hate around you and to respond to it effectively. And your curiosity itself serves as a positive model for others to follow.

## Show Strength

It's not enough to react in the moment to an expression of hate. We must also take positive action to prevent *future* acts from materializing. We must stand up as allies and advocates for others, whether we know them or not and whether they're members of our identity group or not.

The importance of taking forceful action even when hate doesn't affect you or your group directly is an essential part of ADL's values. Our organization was founded in 1913 in the wake of the brutal lynching of Leo Frank, a Jewish man wrongfully convicted of raping and murdering a girl in Georgia. And yet, when the founders wrote a mission statement for the organization, they imagined that the ADL wouldn't just fight antisemitism but would "secure justice and fair treatment to all."

This kind of intersectional commitment might not raise eyebrows today, but it was a bold and audacious claim at the turn of the twentieth century when American Jews were marginalized in many aspects of public life. Even today, the charters of many prominent civil rights organization very reasonably are focused on fighting the ills that affect their own communities. And yet it was a quintessentially Jewish idea for the ADL's founders to seek something greater than their own self-preservation and fight for the rights of others as much as themselves, evoking the ancient sage Hillel's timeless call: "If I am not for myself, who will be for me? If I am only for myself, what am I?"

Looking back on the ADL's history, I'm proud that we've long stood up against all kinds of hatred, not just antisemitism. In 1948, the ADL filed its first amicus brief as part of *Shelley v. Kraemer,* a landmark Supreme Court case that overturned restrictive housing covenants that discriminated against African-Americans and American Jews. In 1953, it filed an amicus brief in the historic *Brown v. Board of Education* case that ended racial segregation. My predecessors marched arm in arm with the Reverend Martin Luther King Jr. during the 1950s and 1960s and advocated against the fearmongering of Senator Joe McCarthy and in favor of the rights of immigrants. It's a little-known fact, but John F. Kennedy's *A Nation of Immigrants,* a book that celebrates America's heritage as a country of immigrants and refugees, was commissioned by the ADL in 1957. And when President Lyndon B. Johnson signed the landmark immigration act into law on Liberty Island on October 3, 1965, ADL national director Ben Epstein stood next to the thirty-sixth president because of his longtime advocacy on immigration issues. I am proud that one of the pens that LBJ used that afternoon hangs in a frame on my office wall alongside the original Western Union telegram inviting Epstein to attend the ceremony.

Inspired by this legacy, I've intensified our commitment in recent years to taking a leadership role in the fight against not just antisemitism but also racism, anti-immigrant xenophobia, and other forms of hatred. In 2016, I was appalled to hear that the incoming Trump administration was contemplating banning Muslims from entering the United States. At the inauguration of the ADL's Never Is Now summit on antisemitism and hate that year, I turned heads by standing up and proclaiming: "If one day in these United States, if one day Muslim-Americans will be forced to register their identities, then that is the day that this proud Jew will register as a Muslim." And I really meant it.

You can gain a sense of how committed the ADL is to fighting hate of all kinds simply by walking into our headquarters. Since my arrival as director, we've mounted in our lobby an enormous photograph from the JFK Library archives of former ADL director Ben Epstein standing next to Reverend Martin Luther King Jr., U.S. attorney general Robert Kennedy, NAACP executive director Roy Wilkins, and President Lyndon Johnson in the Rose Garden at a White House ceremony.

And yet, we haven't always been perfect. For many years, the ADL refused to describe the slaughter of Armenians in Turkey in the early twentieth century as a genocide. That generation of ADL leadership would not do so for several reasons, including a concern that such a reference could diminish the sense of uniqueness of the Holocaust perpetrated against the Jews by the Germans. Previous leaders also didn't want to jeopardize the strong alliance between Turkey and Israel that facilitated the flow of Jewish refugees escaping Iran and other countries in the region. While those issues might have been valid at a certain point, I felt they had blinded us to our moral obligation to call out hate everywhere.

Frustrated by this inconsistency, I moved to rectify it. After consulting with our board and other advisers, I penned an op-ed publicly calling what happened to the Armenians a genocide and taking responsibility on behalf of the ADL for failing to make this statement earlier. This admission drew the ire of Turkish officials. The Turkish media, which routinely espoused antisemitic conspiracy theories, lashed out at ADL. I also incurred the wrath of that country's Jewish leaders, who felt betrayed by my position. One member of the Turkish Jewish community likened me to Brutus but added that at least he'd stabbed Caesar in the back rather than discourteously in his face. At one point, a Turkish nationalist even stalked me in midtown Manhattan and confronted me angrily on the sidewalk. It was an unnerving moment that nearly ended in a fistfight.

Notwithstanding this lapse on the ADL's part, I'm happy to say that the organization's commitment to ending hate of *all* kinds has been far more than just talk or gestures. By seeking justice not just for our group but for all people, we affirm universal principles of decency and compassion. We would want others to stand up for us if we were threatened, so all of us need to stand up for everyone else.

Acting as an ally can take different forms. As noted earlier, it entails supporting others in the moment when they become victims of harassment or bullying. Making sure they're okay. Speaking up to bullies and making them stop. Refraining from participating in the hateful behavior. It entails sending a clear message: abusing people because of what they believe, where they're from, how they look, or who they love is never acceptable.

## Being an Ally: Questions to Ask

*Some opportunities to function as an ally might seem readily apparent, but there might be more need for allyship in your daily life than you think. Take a moment to consider the following questions.*

(1) What problem behaviors are you seeing in your daily life, whether they're directed at you, people you know, or strangers?

(2) What stereotypes, biases, or other beliefs motivated the problem behaviors?

(3) What actions could you take in the moment to serve as an ally?

(4) What specific words could you use in the moment to serve as an ally?

(5) What actions could you take following the incident to serve as an ally?

Allyship also entails advocating for others, taking actions big and small. Here are a few ways to contribute:

- Start a public-awareness campaign.
- Write a letter to your local elected officials demanding that they adopt anti-hate policies.
- Become active online (create hashtags on social media in defense

of a marginalized group, flag noxious content, post material that counters intolerant speech, applaud positive messages, and so on).

- Volunteer with or fundraise for local and national organizations dedicated to fighting bias and discrimination.
- Meet with local officials and their aides to convey your concerns about hate.
- Attend town-hall meetings to communicate your views.
- Rally businesses and consumers to take a stand against individuals and organizations that spread stereotypes.
- Create events at your school or university to educate and inform about hate.

Showing strength usually doesn't mean trying to cancel an offender, as I did with Tucker Carlson. As we've seen, that's an extreme option best reserved for individuals with a long track record of bigotry who refuse to take any responsibility for their actions. Showing strength isn't about vengefully taking someone down because you're angry and appalled. It's about allowing ample opportunity for good-hearted people to make amends even as you stand up and fight for what's right.

## Quelling Your Own Biases

As important as it is to push back against hate perpetrated by others, we can also contribute to a healthier society by getting a better handle on our own biases, including those that might be invisible to us. Many of us harbor *implicit biases,* which the ADL defines as "unconscious attitudes, stereotypes and unintentional actions (positive or negative) toward members of a group merely because of their membership in that group." Implicit biases might lead people to harm others without realizing or intending it — quite different from explicit biases, which people are aware of and that lead to purposeful actions on their part. In practice, the line between explicit and implicit bias is hazy, and the impact of the two might well feel similar to the individuals targeted. Since ultimately it's the impact that matters most, all of us must take steps to address both types of prejudice.

Implicit biases are pervasive in society. For instance, studies have suggested that people tend to view Black men more negatively than they do white men. As the lead researcher in one set of studies reported, after looking at photos of Black and white men, "participants judged the black men to be larger, stronger and more muscular than the white men, even though they were actually the same size. Participants also believed that the black men were more capable of causing harm in a hypothetical altercation and, troublingly, that police would be more justified in using force to subdue them, even if the men were unarmed." Black study participants didn't perceive young Black men as more capable of causing harm or more likely to merit the use of force, but they did perceive them as larger physically than young white men.

To minimize or neutralize our implicit biases, all of us can take steps to become more conscious of them. Educate yourself about other groups. Take time to consider your own biases and how you might have internalized them. Take a fresh look at how you live your life and whether your behaviors might inadvertently be supporting prejudices in society. Ask yourself the following questions:

- What kinds of language do you use in your daily conversations? Might any of it inadvertently degrade or hurt others?
- Do you sometimes stereotype certain types of people or make sweeping generalizations about identity?
- Do you pay equal attention to everyone you meet, regardless of who they are?
- When you organize activities in a professional or leisure context,

do you do it in inclusive ways (for instance, choosing times and places that will allow everyone to attend and keeping costs in check to make participation widely accessible)?

- When people approach you and suggest that something you said or did was culturally insensitive, do you welcome the feedback and take action based on it?

- When you see that the images on the walls of your child's classroom or your workplace lack diversity, do you seek to remedy the situation?

By paying attention to these questions, you can become far more aware of ways in which you might inadvertently help to perpetuate stereotypes despite your best intentions. Check in with yourself periodically; make this kind of self-analysis a habit. Implicit biases are deeply ingrained, and all of us have our blind spots. By constantly pushing into them, we can make the world around us just a little safer and more welcoming for others.

## Coming Together Against Hate

How would you respond if you learned a band of neo-Nazis planned to march in your town? Would you sit by idly? Or would you *do* something?

In April 2018, authorities in Newnan, Georgia, granted a notorious neo-Nazi group, the National Socialist Movement (NSM), a permit for a public rally held to commemorate Adolf Hitler's birthday. Many residents of this small city outside of Atlanta were repulsed by the NSM's vile ideology, and they also feared a repeat of the violence that had taken place at the infamous Unite the Right rally in Charlottesville the previous August. Anticipating the national attention that the NSM's march would draw, they were appalled that their normally peaceful town would become known as a place for hate.

One local citizen, Russell Campbell, decided that he needed to take a public stand against hatred and bigotry. Recognizing that a show of force by the police would likely keep protesters and counterprotesters apart, preventing him from personally delivering a countermessage to the NSM, he had an idea: Why not deliver a message of inclusion and peace from thousands of feet in the air?

Campbell decided that he would hire a plane to fly a banner over the city with a message reading "Newnan believes in love for all." He started a campaign on GoFundMe, hoping to raise $1,200. In making his pitch, he wrote: "We'd really appreciate you helping us counter the message of hatred this unwanted group is bringing to our town with a message of love and inclusivity." In the end, sixty-one people donated over $1,500 in support of the endeavor.

On the day of the rally, the plane flew high and proud over Newnan, adding to the scene as counterprotesters massed. Although technical issues with its engine apparently forced the plane to land early, before the NSM showed up, Campbell accomplished something important: He managed to mobilize dozens of others to take a public stand against hate. Since the flight had cost less than anticipated, Campbell donated the sum that remained—over six hundred dollars—to a local African-American museum and the ADL. "That's not a world-changing amount of money," he told a local newspaper, "but it shows something good coming from the hate the Nazis tried to bring to our town. It shows that people will stand up to protest and put their money where their mouth is to combat that type of hate."

Campbell and his funders hardly were alone in taking on the Nazis. Across the whole town, public opposition was fierce. In the weeks leading up to the march, tens of thousands of people signed an online petition against holding it. Local groups and merchants banded together to create #Newnanstrong, a grassroots community event held the evening before the march that affirmed the community's opposition to hate. "We want there to be a different narrative," an organizer said. "We're a group of people who work toward common goals and we just want that to be the message of Newnan and who we are."

Taking to social media, organizers of #Newnanstrong urged residents to show up and enjoy a range of musical acts and patronize local businesses that for safety's sake would close during the neo-Nazi march. A local arts group asked children to decorate the city park with rainbows and other images of peace, wisely noting, "It will be hard for the hate group to take serious video footage when a rainbow-colored unicorn is in the shot."

On the day of the march itself, hundreds of counterprotesters showed up to oppose the neo-Nazis, shouting "Newnan strong" and carrying signs with messages like "Love Thy Neighbor" and "Hate-Free City." More community members showed up at a special "Interfaith Community Unity Service" held at a nearby church. A number of users posted messages on social media expressing support for the counterprotesters and condemning hate. Billboards trumpeted #Newnanstrong and "a community standing together."

The local newspaper pronounced "We are #newnanstrong" and noted that "the majority of this community abhors the message of the NSM and everything it stands for." Striking a tone of moderation, it also came out against Antifa's tactics of direct confrontation. "Those who want to fight fascism and build better relationships with people don't do it by screaming into the faces of their enemies. They do it by making consistent strides toward improving their communities and investing their time and effort on a continual basis."

I tell you about Newnan to make a simple point. We are not powerless against hate. Each one of us, acting alone, can push back against hate in our daily lives wherever we see it. And when we come together, we can inoculate entire communities against the virus of hate, safeguarding future generations from the prospect of genocide.

In Newnan, the NSM hoped to make a statement and win new aco-lytes. In the end, only a couple dozen of them showed up — a laughable and pathetic display. By comparison, thousands of residents of Newnan came together to make their rejection loud and unmistakable. Blacks and whites. Children and adults. Christians and non-Christians. Each contributed to a message of peace, inclusion, and harmony. And as a result, Newnan is a safer, more peaceful, more humane place.

As effective as we can all be in mobilizing our communities, the problem of hate is too big for individuals to handle alone. We need to rally entire sectors of society to fight against violent extremism and the ideology and prejudices that support it. The next chapter explores how government at all levels can help, not only by passing strong anti-hate policies but by setting powerful norms of civility and respect.

# 8

# MOBILIZING GOVERNMENT AGAINST HATE

The Unite the Right rally in August 2017 sent shock waves through-out the United States and beyond. To many, it seemed as if the awful demons of twentieth-century hate were rising again as a political force. At ADL, we were repulsed, dismayed, and terrified, but we were hardly surprised. We'd been following online chatter among the so-called alt-right about a large white-supremacist event to take place in Charlottes-ville, Virginia, for months. We had alerted local and state authorities about the possibility of violence and had posted public warnings about it the week before the rally.

As it turned out, the event was even bigger and more momentous than we'd anticipated. It emboldened hard-core white nationalists, op-portunistic racists, and others on the Far Right in unprecedented ways. No longer did these extremists need to hover in the shadows, fearful of expressing their views. After mustering the largest demonstration of white supremacists in more than a decade, they had proven to them-selves that they could claim a public voice. "We are stepping off the in-ternet in a big way," one white supremacist told a reporter at the time. The haters also found President Trump's equivocating public response —his claim that there were "very fine people on both sides"—deeply validating. The statement amounted to a stunning departure from the moral certitude that America had come to expect from commanders in chief in the face of indisputable bigotry. For the first time in recent memory, it seemed, white supremacists had a friend and perhaps even an advocate in the White House.

But as the months passed, it turned out that Charlottesville wasn't a total win for the extremists. Citizen journalists and the press corps outed many of the rally's participants, who in turn were ostracized and suffered professional consequences. Fractures emerged in the extremist right, with older, die-hard neo-Nazis at odds with a younger generation of haters who sought to give white-supremacist views a softer public image. Although some celebrated the "collapse" of the so-called alt-right, my ADL colleague George Selim put it more accurately, observing that "America's white supremacist movement is still less in a state of defeat as a state of regrouping—the danger of which cannot be overstated."

Most important, Unite the Right powerfully galvanized public opposition to Far Right extremism in its various forms. As Virginia senator Tim Kaine said in 2018, "We have seen a huge number of people saying, 'I'm off the sidelines now. I'm going to get involved to show that no matter what the president says, no matter what anybody says, we're not a nation of division.'"

One of those people was apparently Joe Biden, who referenced the Unite the Right rally when he announced he would run for president in 2020, claiming that he was fighting a "battle for the soul of this nation." He used the phrase in an opinion piece published weeks after Charlottesville, exhorting Americans to show courage, stand up against hate, and "uphold America's values." On the campaign trail, he further explained that he had not planned to run for the White House but had resolved to do so upon witnessing the ugliness of the Charlottesville rally as well as Trump's divisive response.

On the local level, many officials and activists stepped up to counter violent extremism. Dramatizing this chapter's theme, they mobilized the power of government to help create a more tolerant, inclusive, and civil society. A notable example is Andy Berke, mayor of Chattanooga from 2013 to 2021. Berke already knew well the devastation hateful ideology could wreak. In 2015, a gunman influenced by Islamist extremism shot up military facilities in Chattanooga, killing five. That murderous attack prompted Berke to become active in the Strong Cities Network, launched in 2015 at the United Nations General Assembly and billed as "the first global network of cities working together to build cohesion and resilience to prevent violent extremism." Berke held discussions with State Department officials about the possibility of creating a global council that would bring an array of stakeholders together to address extremism and hate in a comprehensive way.

Now, in the wake of Charlottesville, Berke felt convinced that his city needed to create a local council that would mobilize citizens at the grassroots level under the auspices of the mayor's office. Polarization and white-identity politics had become more palpable in his city, but as Berke relates, it was Charlottesville that really spurred him to action. "I felt the need to do something," he says, "because I never thought that I would see people marching down an American city with tiki torches and have that be condoned at the highest levels of government."

Bringing together a bipartisan group that included religious leaders from all faiths as well as educators, activists, and police, Berke charged the Mayor's Council Against Hate with exploring the factors behind extremism's spread and determining how the community could best counter it. As Berke remembers, he had three basic goals in mind. First, the council would suggest specific policy changes to address hate in Chattanooga. Second, it would serve as a platform for drawing attention to hate and pushing back on it. Third, it would provide resources for local educators struggling to deal with outbreaks of hate. Ultimately, he saw the council as having symbolic value for his city. "It's important for communities to define themselves," he says. "By establishing a Council Against Hate, by making it part of the government, it normalizes inclusivity and togetherness. That was the most important thing to me, as it confers a long-term benefit."

As of early 2021, the council had collaborated with the ADL and other partners to develop white papers, respond vocally to local instances of intolerance, provide new opportunities to report hate crimes, support a podcast about extremism, and conduct antibias trainings. The council had also helped drive specific policy changes. For instance, the city adopted a new policy mandating that every potential hate crime automatically resulted in not just a police report but a follow-up inquiry by a police investigator. It was a small step, to be sure, but it upped the ante against bigots and sent the message that the city took hate crimes seriously.

Reflecting on these activities, Berke believes the council has already made an impact despite its brief existence. When he speaks to members of the local community, he finds that many have become more aware of the need for inclusivity. While much more remains to be done, they feel grateful and reassured that Chattanooga is taking meaningful action. "We needed to say, 'This is who we are,'" Berke reflects. "We needed to say that we were against hate. And we did."

## Taking Action Close to Home

Does your city, county, or town have a council against hate? What else is it doing to counter the epidemic of violent extremism? Our local governments touch our daily lives in so many ways — educating our kids, picking up our trash, responding to our emergencies, keeping our roads clean, and so on. Given their direct impact, they can play a pivotal role in the fight against hate as long as we effectively mobilize them.

The specific ways your local government might make your community safer and more inclusive are many. Policies and initiatives might include these:

- Holding weeklong anti-hate events—like Berkeley, California, did—to raise awareness and rally the community to take up the fight against hate.
- Passing a municipal hate-crime law—like Denver, Colorado, did—that addresses gaps in state-level laws, allowing for better prosecution of these crimes.
- Plastering anti-hate messages on city vehicles—like Gresham, Oregon, did—that reaffirm inclusivity and tolerance as a community value.
- Creating a Compassionate City Initiative—like Westland, Michigan, did—to build "good will among citizens in a way that transcends race, religion, creed, and economic differences."
- Fostering more inclusivity and acceptance of immigrants—like Dallas, Texas, did—by raising awareness about immigration's benefits.

Each of these initiatives has been recognized by the United States Conference of Mayors as a "best practice" that other communities might emulate to foster more inclusion and lessen hate.

If you're a local leader or simply a concerned citizen, invent a best practice of your own. What kinds of specific policies, programs, events, or initiatives might you create to protect your community against hate and render it more inclusive and harmonious? What about introducing antibias education programs in your local school? Or launching an anti-hate book club through the local library? Consider the cultural and social resources you and your community have at your disposal as well as the specific needs or challenges your community has relative to hate. Don't shoulder the burden alone. Bring your neighbors together to generate ideas and create a more civil and compassionate community for everyone.

## Fighting Hate at the State and Federal Levels

As important as local action is, leaders at the state and federal levels must also adopt measures to push back against extremism and foster more tolerance. ADL has advanced a comprehensive seven-step strategy that we call our PROTECT Plan for fighting domestic extremism and organized hate. The point isn't for government to do everything itself. We need a "whole of society" strategy to push back on the poison of prejudice. But government still can and must play a central role.

## The PROTECT Plan

1) **P**rioritize preventing and countering domestic terrorism
2) **R**esource according to the threat
3) **O**ppose extremists in government service
4) **T**ake domestic-terrorism-prevention measures
5) **E**nd the complicity of social media in facilitating extremism
6) **C**reate an independent clearinghouse for online extremism content
7) **T**arget foreign white-supremacist terrorist groups

First, the *P* in PROTECT: The government should *prioritize the fight against hate.* Truly, we must get serious, and the way to do that is to coordinate between federal agencies (which are usually quite fragmented) and between the various levels of government. Is state and local law enforcement collecting and reporting hate crimes to the FBI?

They need to be. More comprehensive, complete hate-crime reporting can deter hate violence and advance relationships between police and local communities.

We also must compel government to coordinate better with other institutions, such as businesses and civil society organizations. Our plan has called on President Biden to hold a summit with representatives from stakeholders across society to gain insight and then create a comprehensive strategy for combating domestic extremism, regardless of its radical dogma or political orientation. Each state, meanwhile, should designate an official to take responsibility for the fight against domestic terror, coordinating across state agencies, law enforcement, and outside partners.

The administration did not hold a "summit," per se, but it held listening sessions with civil society and civil rights organizations. It then released, for the first time ever, a National Strategy for Countering Domestic Terrorism, one that tracks closely with our PROTECT plan. For their part, states could start by convening commissions of experts to better understand how to orient themselves around the modern threat, as the State of Maryland did with ADL's encouragement.

The *R* in PROTECT stands for *resources*—government should **provide sufficient, proportionate resources** for addressing the issue. If we agree that hate threatens the very foundation of our shared society, we must fund a meaningful response. For starters, ADL advocates for the Domestic Terrorism Prevention Act (DTPA), which would require law enforcement agencies to report on domestic threats and authorize relevant government offices to address it. As part of this proposed legislation, Congress should fund these departments and drive those dollars down the line to local actors. Crucially, the resources provided to these entities should be proportionate to the threats on which they transparently report, ensuring that powerful counterterrorism resources address violent threats and not political ideologies. The federal government has boosted resources to fight violent extremism but not by nearly enough to meet current threats. State governments should pass acts similar to the DTPA, making resources available and reporting annually on the state of the domestic-terrorism threat in their locality.

O is for *oppose*—we need to **oppose and root out extremists in public service.** For years now, white supremacists in the United States and Europe and Islamist extremists in the Middle East have made a point of recruiting active-duty and retired members of law enforcement and the military. Extremists often seek out these recruits because they possess invaluable expertise in areas like weapons and urban combat and because they can imbue extremist groups with an aura of respectability. But individuals with extremist ties should be prohibited from serving in law enforcement or the military. This could be done by conducting basic background checks and reviewing social media profiles when making hiring decisions—not to evaluate applicants' beliefs but to ensure that they are not incapable of performing their duties because of their alignment with the criminals they are charged with catching.

When agencies dismiss personnel from service for extremist activity, they should share that information across the government. Sadly, we've seen instances where extremists fired from one department land jobs at another agency. At the federal level, DHS, the Department of Justice, and the Department of Defense have all announced concrete steps they are taking to make sure that extremists are not among those defending our country. We can only hope that they succeed in that mission and that state and local entities will follow suit.

As scary as it is to think of extremists serving as police officers or soldiers, some of them have sought to legitimize their causes by running for elected office. Historians might trace this trend back to Newt Gingrich, who reshaped Washington, DC, with his incendiary politics. The former history professor came to Congress in 1979 as a firebrand freshman legislator and eventually rose to become Speaker. Perhaps best known for his singular role in destroying the bipartisan spirit that had enabled Congress to function effectively for decades, Gingrich urged young Republicans to wage a "war for power" and to "raise hell."

Then there was former KKK leader David Duke's audacious run for a Senate seat in Louisiana in 1990. Derided by a broad expanse of GOP leadership, including President George H. W. Bush, Duke still garnered headlines that earned him a large bloc of votes and established him as a notorious figure on the national stage. In recent years, the rise of celebrity culture, the fragmentation of traditional party structures, and the general mainstreaming of extremism has made it even more viable for extremists to run for elected office. In this context, the ascension of a political neophyte like Marjorie Taylor Greene should come as no surprise, and we likely will see others from all sides try to take control.

The ADL has long pushed back against the efforts of extremists to capture elected office, and we've admonished political leaders to assert some moral authority when they actually break through. Republican majority leader Kevin McCarthy heeded our call in 2019 and removed Steve King, a serial racist who routinely derided minorities for years, from his committee assignments. This enfeebled King, leaving him without legislative power to convert ideas into policy, let alone deliver value for his constituents. Unfortunately, GOP leadership refused to do the same to Marjorie Taylor Greene despite pleas from ADL and other groups, all of whom argued that Greene's terrifying embrace of dangerous conspiracy theories justified limiting her influence on the nation's lawmaking process. The House of Representatives was forced to intervene, and she was stripped of her committee assignments, a rebuke that seemed more than appropriate, given her rhetoric.

Our demands in 2019 that House Speaker Nancy Pelosi censure Rep. Ilhan Omar for making blatantly antisemitic comments about Jews and money went unheeded. While Representative Omar apologized for those remarks, they seemed to foreshadow her comments in 2021 likening Israel to Hamas, an almost profane comparison when you consider that one is a despotic terror organization and the other is an open, pluralistic democracy. Some defended Representative Omar by noting she had compared the United States to the Taliban in the same breath, but one still must consider how her comments were heard by Jews, a minority that perennially faces antisemitism.

We constantly compel people to demonstrate sensitivity when it comes to discussions of vulnerable minorities such as Jews, African-Americans, AAPIs, Muslims, and so forth; this situation should be no different. Of course, instead of ex post facto maneuvers, party officials would have done better to implement an anti-hate mandate well before such people were ever elected, clarifying that hateful statements would not be tolerated and could cost them significantly.

The first *T* in PROTECT means that federal and state governments must *take a variety of measures to prevent extremism*. There is so much that other actors in civil society — universities, schools, nonprofits, and so on — can and must do at all levels to help people reject extremism and intolerance. But they need guidance, oversight, and funding from government. It's vitally important that government fund antibias and civics education in schools and provide for counseling and off-ramping of those who have fallen under the sway of extremists. Public health–style approaches, such as behavioral health and public-messaging initiatives, can help prevent people from becoming extremists, but we should keep these programs separate and distinct from law enforcement processes. The federal government has experimented with such programs but so far has failed to scale them. Individual states have only begun to explore these programs' potential. Congress should also pass legislation that mandates Holocaust education and genocide awareness in the classroom, trains community-based organizations in local and

state hate-crime prevention, implements restorative justice to rehabilitate perpetrators rather than simply punishing them, ensures adequate
data collection about hate crimes, and so on.

The *E* in PROTECT refers to the need for government to take steps
to *end social media's facilitation of extremism.* The world is climbing out
of the rubble of the global pandemic, but the "infodemic" appears to
be an equally insidious and potentially more persistent virus. From pogroms in India catalyzed by false conspiracies spread via WhatsApp to
anti-Jewish memes circulated through Instagram to the relentless normalization of racism on Facebook, it's impossible to ignore the human
toll exacted by hate. And I refer here not just to obvious forms of animus but disinformation (false information that is spread with the intent
to mislead people) and misinformation (false information that is spread
without the intent to mislead). We've already reviewed how social media companies appear almost constitutionally incapable of rooting out
all of these viral variants, essentially serving as super-spreaders of hate.
Government must step in to hold the companies who publish this poison accountable for their actions.

Accountability can take different forms in different jurisdictions depending on prevailing policy. In the United States, we need a drastic
overhaul of Section 230, as discussed previously. While the protection
provided by this provision has yielded rich reservoirs of user-generated
content on sites like Wikipedia and YouTube, it has also shielded companies from responsibility in ways that defy both logic and common
sense. We should strive to carve out a middle path, upholding free expression while still managing the problem of online hate. Social media
services must abide by a set of norms like other companies in nearly all
other industries do. If that means slower growth or trimmed profits, so
be it.

Beyond ensuring the platforms' basic accountability, government
must take additional steps to protect the public online. How much bad

content exists on social media services at any given time? It's impossible to say for sure, which is why we need government to mandate the transparent tracking of online hate and disinformation. Government also should move to provide people who are harassed online with more effective legal remedies. We should legally define cybercrimes like doxing (posting people's personal information online to enable or encourage others to stalk or otherwise victimize them) and swatting (harassing people by calling authorities and falsely reporting an emergency, prompting police or others to arrive at their homes prepared to use lethal force). Once we do this, it will be critical to train police officers to enforce these laws and to adjust our legal systems to discourage such acts via education and enhanced penalties.

The *C* in PROTECT calls for the government to *create an independent clearinghouse for online extremist content.* We desperately need a publicly funded independent center that tracks extremist messaging online while also preserving civil liberties. If there were some form of centralized gathering and analysis of information, we'd be able to search for indications of imminent violence. Structuring the clearinghouse as an independent entity would allow experts to find potentially violent content in real time and provide tips to governments and industry so they could address possible criminal activity. The National Center for Missing and Exploited Children operates in this way, seeking out exploitative content independently from government but providing referrals to law enforcement when criminality is suspected, allowing authorities to quickly identify and protect exploited children. A national clearinghouse that brings together information about extremists' iconography, language, and memes is long overdue.

And finally, the last *T* in PROTECT means that the government must *target foreign white-supremacist terrorist groups.* As we've seen over the decades with Islamist organizations such as al-Qaeda of the Arabian Peninsula and the Islamic State (ISIS), extremist groups often forge global networks. The internet and social media have made this process eminently easier, connecting people across continents and cultures who previously would never have found one another. More recently, right-wing extremist groups have formed international networks; European white supremacists marched at the Unite the Right rally, while American white supremacists have stood shoulder to shoulder at racist rallies in Central and Eastern Europe in recent years. Former White House official Steve Bannon toured Western Europe in 2020 in a thinly veiled effort to recruit like-minded, aspiring candidates who could replicate his "success" at the ballot box with a mix of nationalism and nativism.

Governments must collaborate to disrupt and dismantle these efforts before they take hold. The State Department is charged with developing a global approach to countering white supremacy. As of now, its leaders are ill-equipped to do so, but they have the tools to mobilize a multilateral approach to what has become a global threat. The National Security Council at the White House should ensure that any foreign extremist groups that meet the criteria for designation as terrorist groups are formally classified as such. This would allow the government to leverage a variety of tools to disrupt such organizations.

By vigorously implementing all of these steps, federal and state governments can clamp down on extremist ideology and prevent it from spreading. Complementing steps taken by local government, they can help keep society from rising further up the Pyramid of Hate.

## Setting the Right Tone

When we think of government's role in the fight against hate, it's tempting to limit our focus to policies. But an equally important area of intervention concerns how leaders comport themselves. Leaders hold enormous sway in our society. They set the tone for public discussions and either strengthen or erode informal norms of thought, speech, and action. To prevent society from progressing further up the pyramid, it's vitally important that political leaders and other authority figures take strong, unambiguous stands against intolerance in all of its forms and publicly advocate for inclusion, civility, and respect. We cannot say it enough: America is no place for hate.

You might wonder if speaking out against hate is somehow a "liberal" or "woke" thing to do. It isn't. In the days after 9/11, as the smoke still lingered over lower Manhattan and as the United States was counting the bodies after the most lethal and coordinated terror attack on the homeland in American history, tensions ran high and American Muslims feared a backlash from their fellow citizens enraged by the attack. And yet President George W. Bush did something remarkable. Along with a delegation of officials from the White House, President Bush very publicly visited the Islamic Center of Washington, DC, a large mosque and communal building. In that space, he made remarks that sharply distinguished between the criminal acts of a few and Islam in general. As he noted, "These acts of violence against innocents violate the fundamental tenets of the Islamic faith. And it's important for my fellow Americans to understand that."

Bush explicitly repudiated those who would perpetrate acts of violence or even simple unkindness against Muslims as revenge for the attacks: "Those who feel like they can intimidate our fellow citizens to take out their anger don't represent the best of America, they represent the worst of humankind, and they should be ashamed of that kind of behavior." In a moment of potential moral confusion for the nation, his clarity affirmed basic values like respect, dignity, and human worth.

Now, that's a true act of leadership—a powerful and profound display of compassion and decency. Sadly, it didn't shield Muslims living in the United States from harassment—Islamophobia and anti-Muslim hate crimes surged after 9/11. Sikhs and people of South Asian descent

were also targeted with hate. But one can only imagine how much
more invective and actual violence might have been unleashed against
Muslims, Sikhs, and others if President Bush had opted to stay silent or
had actively encouraged such xenophobic sentiment.

In stark contrast, President Trump's equivocal comments after Charlottesville (the unforgettable line there "were very fine people on both
sides") and in the wake of the murder of George Floyd (he expressed
sadness about Floyd's death but also derided protesters as "thugs" and
ominously tweeted, "When the looting starts, the shooting starts")
sent a very different message. This is to say nothing of the innumera-
ble other racist comments he made and dog whistles he sent during his
presidency, such as planning to mark the African-American holiday of
Juneteenth by holding a rally at the site of an historic racist atrocity.

The meaning of Trump's behavior wasn't lost on anyone, certainly
not on the extremists. After Trump angrily tweeted that Congresswoman Omar and three other women of color serving in Congress
should "go back and help fix the totally broken and crime infested
places from which they came," the prominent neo-Nazi Andrew An-
glin exclaimed, "Man, President Trump's Twitter account has been
pure fire lately. This might be the funniest thing he's ever tweeted. This
is the kind of WHITE NATIONALISM we elected him for." When a
political leader, especially one with the tremendous reach and influence
of President Trump, not only fails to condemn hate but actively encourages it, the entire tone of our nation's discourse shifts. Extremists
become emboldened, hateful ideology spreads, and society becomes far
more susceptible to violence.

Elected officials at every level of government bear responsibility
for calling out bigotry at every opportunity and for refraining from
supporting extremism themselves. When they shirk this responsibility, we must put pressure on them, regardless of their political party.
As previously mentioned, when Representative Omar tweeted in

early 2019 that Jewish money had translated into support for Israel, affirming the classic antisemitic notion that Jews are wealthy, malicious manipulators of events, I demanded that leaders of the House of Representatives take action, noting that "we expect our politicians to condemn bigotry, not to fuel it." She apologized, but unfortunately she has made a habit of such behavior, spewing bigotry and then apologizing on multiple occasions. In 2021 we saw Rep. Betty McCollum accuse Israel of "bombing Gaza into oblivion," a blatant and irresponsible mischaracterization of the country's response to a wave of Hamas rocket attacks. Such comments seemed bound to stir up anger and cause violence rather than shed light on a complex conflict initiated by a terrorist group. Considering the anti-Jewish violence that preceded her inflammatory exaggeration, it seems more than reasonable to question her motives.

Of course, stepping up and speaking out is easy when you're pointing out the failings of a perceived opponent. It's far more difficult to call out members of one's own tribe, especially in our polarized society. But we must be willing to analyze statements and beliefs on their own merit, regardless of the speaker. And we must be honest about our *own* positions to ensure that they don't veer into hate. We can't allow a hierarchy in which members of one's own group are more protected from hate — and excused from hating — than members of other groups. We must be willing to call out prejudice on our side too.

My friend Natan Sharansky, a former prisoner of conscience from the Soviet Union, impressed this lesson upon me most memorably during a November 2017 conversation we had in San Francisco. For the simple crime of expressing his Zionist beliefs, Sharansky was imprisoned in a gulag and then confined to his home for years, yet he never lost his dignity. His personal struggle for liberation symbolized the communal struggle of Soviet Jews and helped trigger a worldwide human rights movement that eventually led to their freedom. After he immigrated to

Israel, the global community embraced him for his moral character, and he remains a revered figure to this day.

During our conversation, held in the wake of Charlottesville when my sense of moral indignation was high and the press routinely noted my willingness to take on President Trump, Natan emphasized his view that calling out hate in others has far more power when you have the courage to call it out among your own friends. If you're a progressive, you can't blind yourself to liberal politicians who spread hate, even if they do it in the name of progressivism. When you speak out against that type of intolerance, you gain far more currency than if you limit yourself to simply taking down someone from the other side. If you're a conservative, the same rule applies. You must call out your ideological fellow travelers who spread hate even if you like them as candidates.

Reflecting on this simple but sterling bit of wisdom, I've come to believe that all elected leaders deserve a seat at the table irrespective of their identities as long as their ideas are grounded in decency and evidence. Let's remember that people with whom we agree also bear responsibility for behaving decently and treating others with respect. And let's remember as well that we can't always rely on leaders to fight hate on everyone's behalf.

We have seen this reality play out on college campuses; anti-Israel activists have hounded Jewish students while university presidents remained silent. Off campus, some left-wing community leaders also fail to call out problematic rhetoric. A prominent anti-Israel activist in New York City, law student Nerdeen Kiswani, publicly stated her support for Hamas actions in spring 2021. She did so even though Hamas is a terror organization that seeks to kill civilians and that promotes a version of Islam that resembles the oppressive, chauvinist model practiced in fundamentalist Iran. Yet leaders in the progressive movement with whom she often partners haven't disavowed her.

Elected leaders in the United States and elsewhere have often fallen silent when their parties' candidates make antisemitic statements and then claim that their opposition is to the Jewish state, not necessarily to the Jewish people. This posture sounds reasonable on the face of it, but in all too many cases it falls apart under the mildest scrutiny. People who say they oppose Israel, not Jews, often play a double game, failing to stand up to defend Jewish people when they are attacked. Ultimately, all of us bear responsibility for speaking out. If we see hate in our midst, we have to say something. Standing up for decency and civility should start with our elected officials — but it can't end there.

## Rediscovering Our Common Connections

When we think of hate, we usually conceive of it as vitriol directed at others because of some immutable characteristic, such as faith, race, gender, or sexual orientation. But in many countries around the world, we've seen a great deal of hate directed across political or ideological lines. And so I believe that we must broaden our definition of hate to recognize that ideology has become a fundamental element of identity in many people's minds, even if it's less discernible than race or religion, and that ideological identity, like these other identities, can serve as a basis for hatred. If we're going to heal societies and prevent a conflagration, we have to address ideologically based hate.

American democracy isn't just the Washington Monument, the Lincoln Memorial, the Capitol Building, and the rest of these massive, neoclassical structures in Washington, DC. At its core, democracy hangs together due to an invisible but resilient filament that binds our society: trust among citizens. We Americans are strong when we believe in one another and work to preserve the country's best interests. When trust frays, you have neighbors fighting neighbors, family members clashing with family members. And trust today is dangerously frayed.

Our political discourse is not merely polarized but riven by outright extremism and a potentially explosive political sectarianism. Many Republicans no longer perceive Democrats as political opponents with whom they disagree; they see them as hardened enemies who have orchestrated an anti-democratic coup against Donald Trump. An increasing number of people regard groups like the Proud Boys and the Three Percenters as legitimate political actors instead of what they truly are: bands of thugs whom we should imprison when they commit acts of violence and otherwise resolutely marginalize.

And sectarianism is by no means unique to the Right. Many Democrats believe that all Republicans are committed to maintaining power at all costs, even if it means sacrificing democracy. Not long ago I heard Rep. Alexandria Ocasio-Cortez tell an audience at the *New Yorker* Festival that bipartisanship was overrated as an outcome and that it hadn't generated progress. She dismissed the impulse to work across the aisle as a "vintage fantasy" and one responsible for various sins, including "the Iraq war . . . and bank bailouts." I could not have been more frustrated. Sure, some bipartisan initiatives have failed, but many more have succeeded over the centuries, making the United States the most durable and vibrant democracy in history. Moreover, the willingness to compromise with others, even those with whom you politically disagree, is crucial to a healthy, functioning society.

With extremists dominating public discourse, a negative dynamic takes shape and produces polarization. Our positions become hardened, and as misinformation about our opponents circulates, antipathy flares. Locked into our filter bubbles, we can't even agree on a shared reality or separate fact from fiction. Dialogue of any kind appears to be a fool's errand, and compromise with the other side is tantamount to surrender. Perceiving our opponents as mortal enemies with whom we have nothing in common prevents any kind of productive engagement, let alone political compromise. We shouldn't be surprised if violence ensues, as it did during the insurrection at the U.S. Capitol. Considering recent history in Northern Ireland or the former Yugoslavia, we should be surprised that such violence doesn't flare up more often.

How do we reduce sectarianism and convince polarized Americans to engage with one another again? It's simple: We must bring people together outside of their political and social media bubbles so they can rediscover their shared humanity and build trust. We can't wait for reconciliation to happen spontaneously. From youth groups to bridge clubs, we must work at it at the grassroots level. But certain policy interventions could also catalyze breakthroughs.

One proven approach to bridging divides is national service. In late 1991, while I was mopping floors and cleaning the cafeteria at Tufts Dining Services to defray the cost of my college tuition, I heard about a young presidential aspirant from Arkansas who was lauding two exciting new nonprofits, Teach for America and City Year. These programs put recent college graduates to work in communities as a way of paying down their student debt. They seemed to me like an amazing concept and a much more productive way to deal with my debt than working part-time in the lunchroom.

After graduating, I moved to Little Rock to work for that candidate, in large part because I was so enamored by national service as a policy solution. Little did I know that Governor Bill Clinton would go on to win the White House or that almost twenty years later I would find myself in a senior role in the Obama White House overseeing domestic national-service programs as director of the Office of Social Innovation and Civic Participation.

These programs are remarkable when it comes to combating hate. Whether you're Black or white, Republican or Democrat, Jewish or Christian, gay or straight, you're working on the ground side by side with others, engaging in tasks like teaching at-risk kids, building houses, or helping out with disaster relief. It's a transformative, life-changing experience, one that builds solidarity between participants and those they serve on a basic human level.

National-service programs are also extremely cost-effective for the government, since nonprofit organizations or donor organizations foot part of the bill. Giving participants a year of national service costs the government less than a year of unemployment insurance. Participants build skills that can be highly relevant in the workforce while also providing the federal government with much-needed labor.

Imagine creating service programs at the local, state, and federal levels funded by public money as well as foundations and corporations. We could repair the civic fabric and rebuild our shared sense of purpose block by block, neighborhood by neighborhood, town by town, city by city, state by state. If we could get kids from red states like Texas and Arkansas working with peers from blue states like Massachusetts and California, we'd have a shot at overcoming the ideological polarization that threatens to destroy us. The point isn't to create a national-service corps dedicated specifically to fighting hate. It's to recognize that national service of all kinds is inherently pro-social and inimical to hate.

Other important programs for combating excessive partisanship are those that convene people across ideological lines to explore complex issues. The Civil Society Fellowship, jointly run by the ADL and the Aspen Institute, was designed exactly for this purpose. It brings together diverse young leaders from across the ideological spectrum for passionate discussions held in a dispassionate manner. As glib as that might sound, these leaders talk together and build relationships with one another, ideally learning to look beyond politics to solve problems. That's right—even if some high-profile members of Congress are cynical about the merits of bipartisanship, I believe it's the mortar that binds together a healthy democracy.

Every year, about two dozen bright young leaders participate in the Civil Society Fellowship (CSF). Admittedly, some Far Left activists have attacked it, perceiving the ADL as an anti-progressive organization. I also have heard from conservatives who refused to take part or even apply because they perceive ADL and Aspen to be liberal institutions. Despite these attacks from both sides, I am excited that a decade from now we will hopefully have hundreds of activists, public servants, faith leaders, and nonprofit professionals across the country who have been through the CSF program and who are committed and capable of reaching across ideological lines.

On the most basic level, we must teach people of all political persuasions to be more pragmatic and intellectually honest. You can be principled and also aware of the limits of your principles. You can be principled and also recognize elements of opposing ideologies that have merit. One need not be a Republican to value the important achievements of Republican administrations, such as the passage of the Clean Water Act and the creation of the Environmental Protection Agency. One need not be a Democrat to appreciate the good that Barack Obama did during the Great Recession by bailing out the automobile industry and saving hundreds of thousands of jobs. We all can get in the habit of questioning our beliefs rather than following them blindly. A resurgence of pragmatism and intellectual honesty would go a long way toward creating a post-hate society.

I think of government action, including programs that foster national service and bipartisan dialogue, as giant vaccination programs for the country. Acting collectively, we can inoculate citizens and communities against hateful ideologies while responding forcefully and effectively when hate crimes do occur. The specter of Charlottesville is painfully fresh in the psyches of many Americans, and that's a good thing. Let's stay motivated, like Chattanooga mayor Andy Berke, to mobilize government and prevent our society from ascending the Pyramid of Hate.

There's another powerful vaccination program we can activate, one whose effects take hold over the long term. I'm talking about education. We must teach young people to recoil from hate and stand up for a safer, more harmonious future. We also must teach them how to think critically and participate as citizens in a democracy. With dedication, care, and patience, we can turn the next generation against hate, keeping them safe from the extremist menace.

# 9

# RAISING HATE-FREE KIDS

Each spring, Jews around the world gather with their families to celebrate their ancestors' historic liberation from bondage and deliverance to the Promised Land. The eight-day Passover holiday begins with two elaborate seders—ritual feasts organized around the telling of the biblical story of redemption—held on successive evenings. These joyous occasions run late into the night and are full of singing, prayers, delicious food, and the mandatory drinking of four cups of wine.

Unfortunately, Passover was anything but joyous for the Bernstein family in the spring of 2019. Twelve-year-old Rachel Bernstein attended sixth grade at a public middle school, one of the largest in the state. Although there were over two thousand students at the ethnically diverse school, she was one of only a handful of Jews. On the day before the first Passover seder, Rachel was riding the bus home from school, sitting alone in the front seat right behind the driver, when an eighth-grade boy near her began shouting antisemitic slurs. "Jews are just awful," he said. "Jews should die. The Holocaust was good. All Jews must die."

Rachel was shocked. Never had she heard anything like this. Without thinking, she turned around and said, "I'm Jewish."

The boy glanced at a friend sitting next to him and then said, "Oh, you should die too. All Jews should die. Jews are horrible people." For several minutes, he continued raging against Jews and advocating violence.

Rachel was terrified. The boy shouting slurs lived in her neighborhood. He was physically bigger than she was and a year older. She was all alone with nobody to protect her. "No other kids on the bus did anything about it," she remembers. "I know they heard it, but they just continued their conversations. One kid was even cheering him on. If something bad had happened, I'm not sure if anyone would have done anything about it."

The bus pulled up to her stop, where her father was waiting to walk her home, as he often did. She was shaking and sobbing as she got off, terrified at the thought of having to ride the bus to school the next day. "It was as if the Earth was crumbling down around me. I'm glad I didn't have to walk home alone from the bus stop that day. So much running through my mind. How would this affect me? Would this kid hurt me tomorrow? I was scared and didn't want to go back to school."

When Rachel's father, Gary, heard what had happened, he was understandably very concerned. The next morning, he drove his daughter to school and e-mailed administrators asking for a thorough investigation of the incident. He wanted to ensure that his daughter would be safe riding the bus home. Although he hoped the school would understand the seriousness of what had happened and respond immediately, Gary heard nothing until midafternoon, when the school safety officer called and told him he didn't think the boy's words on the bus were a threat or danger. He observed that the boy had referred to Jews as a group, not to his daughter specifically. He contended that this somehow made the threat of violence less real. He was a good boy, the officer said, and didn't mean any real harm.

Gary found this response baffling, to say the least. Given that school shootings had become almost commonplace, he couldn't understand how the officer could view the incident so cavalierly. From what his daughter had said, the boy sounded deadly serious when making his threats. Would he beat her up? Would he return with a weapon and assault his daughter, or worse? The school's assistant and head principals both echoed the officer's response, brushing off the incident and telling Gary that the perpetrator was a good boy. As nobody at the school could assure him that the boy wouldn't be sitting behind his daughter again on her ride home, Gary left work early that day and picked her up himself.

For days and then weeks, Gary continued to drive his daughter to and from school, concerned about the environment on the bus. Gary demanded that the school take a series of reasonable steps, such as removing the offender from the bus for the rest of the year, speaking out formally against antisemitism, keeping a watchful eye on the offender to ensure that he wasn't a threat, and involving law enforcement. The principal and other school officials continued to minimize the incident. Gary became concerned not merely about continuing threats to his daughter's safety but about the message the school's inaction was sending. "Rather than seeing a swift and decisive response," he says, "the other students saw the opposite. It was almost like Rachel was punished and off the bus, and he was back on, still running his mouth on the way home."

Sadly, Rachel's terrifying experience is hardly unique. In a 2017–2018 survey of five hundred high-school principals, more than 80 percent affirmed that students at their schools had spoken disparagingly about members of specific groups. A separate survey of over twenty-seven hundred educators found that two-thirds of them had "witnessed a hate or bias incident in their school during the fall of 2018." Bullying rates at schools have remained disturbingly high over the past decade; between 2011 and 2019, about one-fifth of all students reported that they'd been bullied.

Despite this rampant incivility, prejudice and bias aren't ingrained in our DNA. They're *learned* attitudes and assumptions that are "carefully taught" to us by our social and cultural environments. Kids discover how to hate from watching movies and television, from listening to music, and from consuming social media. They absorb it from their peers and from observing the behaviors of adults in their lives. But the culturally constructed nature of our social perceptions means we can also teach kids *not* to hate. We can intervene actively with them from a young age, acknowledging our privilege where appropriate, modeling inclusivity and tolerance, teaching them to value diversity and difference, and helping them become critical thinkers and involved citizens.

To keep our society from rising up the pyramid, it's imperative that we step up at home and at school and emphasize the teaching of inclusivity and tolerance. The best way to stop the rising tide of intolerance is to interrupt hate *before* it takes root. At home and at school, we must drum *positive* messages about others into our impressionable children, instilling strong, humanitarian values that will override the negative messages they'll encounter in the wider world. We must prepare kids to stare hate in the face and roundly reject it.

## Talk to Your Kids About Hate

Let's start with how parents can fight hate at home. Research has found that kids can adopt racist ideas when they are as young as three years old, and the learning process doesn't even take very long—days, not weeks. It thus becomes important to actively teach inclusion and tolerance in the home starting when kids are very small. If we neglect bigotry and pretend it doesn't exist, we leave our kids more vulnerable to hateful voices. We have great power to protect our kids, but only if we make educating about hate a priority.

It's especially important to address hate with our kids given the general lack of control we have over what they see and hear. When I was young, media was far less fragmented than it is today. I remember coming home from school and having only a few television channels from which to choose. The fare they offered consisted mostly of reruns of old sitcoms like the *Brady Bunch* and *Gilligan's Island*. In the evenings, my parents and brother and I watched shows together, as did many families.

Today, it is a far different story. Most middle-schoolers have phones that continuously pummel them with messages, both positive and negative, from a dizzying range of sources. TikTok has replaced TV. Spotify has replaced the stereo. Even the most vigilant parents find it a challenge to monitor much of what their children see and hear, let alone make sense of it.

But we can't just leave our kids to their devices and trust they'll develop positive values. Rather, we need to talk with our kids about hate, covering a range of topics related to identity, prejudice, stereotypes, disinformation, and so on. It sounds so simple, but many of us neglect to do it.

Robert Trestan is the regional director of the ADL's New England office; he oversees educational efforts in schools across five states. As he observes, "Sometimes parents are scared to bring up the hard topics, or they might not realize how much their kids already know about what's going on. They also don't always understand the importance of holding these conversations in the home, in safe settings, and how it prevents [their kids] from believing everything they see online and taking it as the controlling narrative."

Having strategies in hand can help alleviate unspoken fears that keep you from discussing difference and hate with your kids. The ADL website offers a wealth of material for parents, including numerous "Table Talk" conversation guides that provide background information on specific hate-related topics, links to additional resources, and lists of suggested questions you can ask to start the conversation and keep it going. ADL's highly experienced team of education professionals developed them, bringing together academic pedigrees, classroom experience, and hands-on knowledge of what works. By way of introduction, I've drawn on some of the best of this material in the pages that follow.

Research suggests that holding regular dinnertime conversations with your children is a great thing to do in general, conferring important intellectual and academic benefits, building stronger family relationships, lowering stress, and more. Our conversation guides can facilitate conversations on an array of current topics, everything from racial slurs to mass shootings, media bias to specific events like the murder of George Floyd, extremist recruiting techniques to Confederate monuments.

As you hold these conversations, determine what your kids know about a given topic and then offer more information. Ask your kids what else they're curious about and work with them to perform more research. Be sure to ask open-ended questions that deepen the conversation. Listen carefully to their responses, resisting the urge to pass judgment. Consider what action your children might take to get involved.

For instance, if you're talking with your high-school-age children about social movements like #MeToo and Black Lives Matter, you might challenge them to explore the history of protest movements in the United States, learning more about a specific march or demonstration and the difference it made in American life. You might encourage them to ask critical questions about the movements themselves as well as the issues, pushing them to develop their own opinions rather than submitting to conventional wisdom or groupthink. You could encourage them to do online research, attend a march on a topic of interest, or even start a group and hold a demonstration related to an issue that matters to them. You also might advise them to follow people on social media who are attending specific protests that resonate with them.

Not every subject will make for suitable dinner-table conversation with younger kids, so you'll obviously want to approach the conversation differently depending on your child's age. If your children are very young, there is quite a bit you can do beyond dinner-table conversations to help nurture a sense of inclusivity. Something as simple as using music or arts projects to expose them to different cultures and languages goes a long way; it helps kids ponder and explore differences and affords you opportunities to broach the subject with them in ways that feel natural and safe. Also, you might proactively remove objects from the home that somehow affirm common prejudices or stereotypes, such as food brands rooted in a racist past or athletic gear that reinforces stereotypes about Native Americans.

One challenge faced by parents who want to teach their children about inclusivity and tolerance is helping kids process violent, hate-related incidents. You might feel tempted to quickly change the subject, recognizing the fear, confusion, and other emotions these events might arouse and feeling unsure exactly what to say. Although such a response on your part is certainly understandable, engaging around these events can provide an opportunity to reinforce your own family's beliefs, values, and traditions.

Here are some tips to keep in mind:

- *Prepare beforehand,* not only familiarizing yourself with the facts but also gauging your own emotions and the factors that might be coloring your perceptions.
- *Stay alert* as you hold the conversation, looking for signs that it might be causing distress to your child. If your child is acting out, becoming withdrawn, or expressing fears explicitly, this might indicate that the conversation is too much for your child to handle right now.
- *Listen carefully* to what your kids are saying, resisting the urge to ignore or silence them. Clarify any questions that your children have, making sure that you're addressing their deeper concerns.

- *Paraphrase your children's statements for them,* confirming that you understand and clarifying their meaning. For example, you might say, "It sounds like you're afraid that something like this might happen to you."
- *Reassure kids that they're safe and that you love them.* Let them know that while some people do things that are harmful and hard to understand, we live in a wonderful country and for the vast majority of the time they are safe.
- *Be honest with your kids in an age-appropriate manner.* If you don't know an answer to one of their questions, say so. Correct yourself if you've given inaccurate information, because you model for children the idea that it's okay to make mistakes and learn from them. Don't give kids more information than they need, and try to avoid conveying a sense of hopelessness while still acknowledging the seriousness of the situation.
- *Take action* in order to teach your children that they're not powerless against hate. Brainstorm ways you and your children might fight hate in your local community. Explain the connection between hateful words and physical incidents and teach your children how they might best react when encountering name-calling and bullying.

As your children grow and mature, check in regularly about their media consumption, alerting them to stereotypes embedded in entertainment. When watching TV or reading with your children, talk to them about portrayals of identity you've encountered and how they affect their preexisting views. Promote empathy for others by posing questions to your children about characters they have encountered (for instance, "What do you think that experience was like for them?," "How do you think they feel?," "What would you do in that situation?," and so on). Help your children become more media-literate by pointing to patterns they might encounter in the representation of difference. How accurate are these representations of others? What biases or assumptions do they reflect? Helping kids become aware of the ways that media represents the world can help them consume entertainment more critically, understand the complexity of identities, and empathize across the chasm of personal difference.

## Creating Inclusive Classrooms

As important as parents are as bulwarks against hate, we must acknowledge the powerful role of schools. Our kids spend most of their waking hours in the classroom, and their classmates and peers exercise an extraordinary amount of influence over them, especially as they become teenagers. As Rachel's story at the beginning of this chapter suggests, bigoted attitudes from the wider world all too often filter in, causing harm and poisoning the learning environment. If we're going to prepare young people to understand and reject hate when they encounter it, we must reinforce values of inclusivity and tolerance in schools and arm staff to respond properly when incidents of hate occur.

We also must teach kids to become active and thoughtful members of society, recognizing that anti-hate and pro-civics go together. Research reveals a striking lack of civics education and knowledge in the United States. As of 2018, only nine states required students to take a full year of civics classes. One survey found that only about a quarter of Americans could name the three branches of government (legislative, judicial, and executive). This lack of knowledge makes our political system vulnerable to forces that would twist democracy into tyranny. It creates space for hate to fester, fostering a lack of responsibility and a laziness when it comes to following public affairs and ascertaining facts.

Civics education can help inoculate society against hate over the long term. The more kids see themselves as equal participants in a democratic society and the more they understand their civic responsibility to debate and discuss issues thoughtfully, the more they'll cast a skeptical eye on disinformation and obvious hate speech. With strong civics education, they'll feel empowered to speak out against oppression when they see it and build the more tolerant and inclusive society they want to see.

Critics on the right might think that I'm advocating for schools to become bastions of woke ideologies or that I'm pushing for specific interventions, such as teaching critical race theory (CRT). The truth is more nuanced. As a white man, I acknowledge the privilege and status that this identity bestows upon me. At the same time, as a Jew, I know the reality of historical oppression and how it has shaped our community across continents and cultures, driving us out of our homes time and again. I believe that we must acknowledge that pain and trauma endured by minority groups in the past lives on through future generations. If we do not recognize and resolve historical pattens in a just and equitable manner, they will inevitably repeat themselves. Institutionalized racism in all its forms has shaped experiences and outcomes across society in ways too numerous to mention. We thus must aim to teach the full history of our nation and achieve a degree of truth and reconciliation, even if it is hard and pushes people to give up some amount of privilege. To the extent that Ibram X. Kendi and other thinkers associated with critical race theory have shed light on embedded and often invisible forms of oppression, they've done us an enormous and important service.

That said, some proponents of CRT and related disciplines go too far, distilling *everything* down to historical dynamics of oppression and shrinking each of us as human beings to little more than contemporary conventions of ethnic or racial identity. Such advocates seem inclined to easily lump all people into the categories of "oppressor" and "oppressed," denying individuals any degree of agency and negating the nuances that shape communities over time. More fundamentally, they seem to verge on a kind of rigid illiberalism, regarding basic concepts such as logic, legality, and truth "as fictions constructed by the 'white cisheteropatriarchy.'" Such reductionism might seem innocuous in certain contexts, but it is harmful when it hardens into conventional wisdom or is taught to impressionable children.

CRT is a valid field of scholarship, but overreaching interpreters of this discipline oversimplify problems and obscure other attributes and issues, focusing excessively on the issue du jour and potentially introducing new forms of bias. As an example, consider a 2016 law in California mandating a new ethnic studies curriculum. In its original form, the law introduced a doctrinaire pedagogy that, among a litany of glaring flaws, espoused a revisionist history of world affairs that erased the historical suffering of select minorities, including Armenians, Sikhs, and Jews. Here, an effort supposedly designed to reduce ignorance actually threatened to perpetuate it. It legitimized an ugly antisemitism that overlooked the multiethnic and multiracial nature of the Jewish community as well as the millennia of marginalization that Jews have suffered and stereotyped Jews simply as "white" or racially privileged. And this is hardly an isolated case. Some activists also openly promote "narratives of greed, appropriation, unmerited privilege, and hidden power—themes painfully reminiscent of familiar anti-Jewish conspiracy theories."

I believe the challenge is to find a path between ignoring present-day inequities and seeing the world purely in an oppressed/oppressor paradigm. That middle ground, as I see it, is a clear focus on antibias, anti-hate, and civics training. Instead of viewing people as either wholly good or wholly bad, let's acknowledge the reality of racism as we help kids recognize their differences and diversity as strengths and show respect for everyone. Let's enable kids to empathize and understand others and refrain from behavior that makes others feel unwelcome or marginalized. And let's encourage children to notice the injustices around them and speak out on behalf of oppressed people.

Addressing hate and bias can feel daunting for educators, especially in today's charged political environment. How should you talk about topics like antisemitism, racial injustice, and nontraditional families in the classroom? How should you address frightening acts of violence when they occur? How should you adjust your teaching to make everyone feel welcome? Some of the principles that work in the home —like asking open-ended questions, listening carefully, and watching for signs of discomfort—apply in the classroom as well. Keeping some additional strategies in mind can give you more confidence addressing hate and fostering values of inclusion and tolerance.

On the broadest level, educators should strive to create what we at ADL call an "anti-bias learning environment" in their classrooms. As important as it is to discuss topics related to bias and hate when news stories about them appear, creating an inclusive, respectful classroom involves ongoing effort and focus. It's also a multidimensional task that requires a range of tactics. If you are an educator, consider the following questions:

1. Have you recently read any books or articles or watched any documentaries to increase your understanding of the particular hopes, needs, and concerns of students and families from the different cultures that make up your school community and beyond?
2. Have you participated in professional development opportunities to enhance your understanding of the complex characteristics of racial, ethnic, and cultural groups in the United States?

3. Do you try to listen with an open mind to all students and colleagues, even when you don't understand their perspectives or agree with what they're saying?

4. Have you taken specific actions to dispel misconceptions, stereotypes, or prejudices that members of one group have about members of another group at your school?

5. Do you strive to avoid actions that might be offensive to members of other groups?

6. Do you discourage patterns of informal discrimination, segregation, or exclusion of members of particular groups from school clubs, committees, and other school activities?

7. Do the curricular content and wall displays in your classroom reflect the experiences and perspectives of the cultural groups that make up the school and its surrounding community?

8. Have you evaluated classroom materials and textbooks to ensure they do not reinforce stereotypes and that they provide fair and appropriate treatment of all groups?

9. Do you use classroom methods, such as cooperative learning, role-playing, and small group discussions, to meet the needs of students' different learning styles?

10. Do students have opportunities to engage in problem-solving groups that address real issues with immediate relevance to their lives?

11. Do you use a range of strategies, in addition to traditional testing methods, to assess student learning?

Working through these questions can help you as an educator achieve a new level of self-awareness about your efforts to fight hate. Even if you feel that you've been sensitive to bias, you'll likely find that there's much more you can do in the classroom and beyond to foster tolerance and inclusivity and create an antibias learning environment. Conversely, a failure to critically assess your teaching from the perspective of antibias can lead to unwittingly supporting bias in the classroom or failing to address it sufficiently when students articulate biased views.

Of course, it's not enough to put your own teaching practices under the microscope. The broader school context also influences the learning environment in individual classrooms. Consider these questions about your school:

1. Does the school's mission statement communicate values of respect and inclusion?

2. Do students typically interact with one another in positive, respectful ways?

3. Do the school's symbols, signs, mascots, and insignias reflect respect for diversity?

4. Do celebrations, festivals, and special events reflect a variety of cultural groups and holidays?

5. Is the school staff (administrative, instructional, counseling, and supportive) representative of the racial, ethnic, and cultural groups in the surrounding community?

6. Are staff or volunteers who are fluent in the languages of families in the school community available?

7. Do students, families, and staff share in the decision-making process for the school?

8. Has the school community collaboratively developed written policies and procedures to address harassment and bullying?

9. Are consequences associated with harassment and bullying policy violations enforced equitably and consistently?

10. Do the instructional materials used in the classroom and available in the school library, including textbooks, supplementary books,

and multimedia resources, reflect the experiences and perspectives of people of diverse backgrounds?

11. Are equitable opportunities for participation in extra- and co-curricular activities made available to students of all genders, abilities, and socioeconomic groups?

12. Do faculty and staff have opportunities for systematic, comprehensive, and continuing professional development designed to increase cultural understanding and promote student safety?

13. Does the school conduct ongoing evaluations of the goals, methods, and instructional materials used in teaching to ensure they reflect the histories, contributions, and perspectives of diverse groups?

As ADL has learned, each of these actions can help create a positive, antibias environment where teachers and students teach, model, and experience respect for diversity. If certain of these areas remain weak or unexplored in your classroom or school, approach them as opportunities for future growth and development. Look to integrate cultural diversity into all aspects of your teaching rather than relegating it to a "cultural history month." Stay abreast of current antibias education issues and discuss them with students, posting articles from newspapers and magazines in the classroom. Let your students know that you're engaging in the learning process too. Stay in touch with your own personal cultural biases and assumptions, as doing so will allow you to model inclusion better for others.

Again, this work is an ongoing process of growth and discovery, one that won't necessarily move linearly. Intolerant thinking certainly might emerge from time to time among students, and we ourselves might occasionally be guilty of it. If some students complain that something you said or did offended them, try to react in a nondefensive way, modeling this behavior for them. Assume goodwill and make that assumption a common practice in the classroom. Affirm that lapses in judgment will happen, recognizing that none of us is perfect.

Getting the tone right makes all the difference in how students respond. A collaborative approach that encourages student participation works best. Try to avoid preaching to students about how they should behave, as that usually backfires. Instead, provide opportunities for students to resolve conflicts, solve problems, work in diverse teams, and think critically about information. Of course, when students purposely commit acts of bias, it's important to intervene. Kids take careful note of what happens when someone becomes a target of hate. Staying quiet in the face of injustice conveys the idea that you're condoning the behavior or at least not recognizing its importance.

The unusual circumstances created by the COVID pandemic posed new challenges and opportunities for educators, and some new tactics like distance-learning and online activities are here to stay. Administrators and teachers need to adjust their approach on multiple fronts, including how they deal with bias and hate. When physical schools are offline, we must work to create virtual classrooms that are safe, respectful, and inclusive. Establish clear norms of behavior for virtual classrooms, supplementing rules that would apply offline with others specific to digital spaces (no side conversations in chat, for instance, and no sharing offensive images). Since it can be harder to gauge students' emotional needs online, we must devote a different level of investment and attention. Institute regular mood and emotional temperature checks where students can share how they're feeling using an emoji, a thumbs-up or a thumbs-down, a 1-to-5 scale, or a one-word feeling check-in. Follow up if there are students who seem overly distressed.

Bias, bullying, and online hate can happen under the radar. Students sometimes send hateful messages using the chat function on videoconferencing or messaging apps. As much as possible, disable specific features or functions that lend themselves to this type of behavior and check in with students regularly to assess whether this is happening. Work with your school or district to understand and improve those guidelines. Reflect on how your practices (for example, discipline, materials, access, routines, participation) promote or diminish equity in your virtual classroom.

My discussion here has touched on the basics of creating an inclusive learning environment that instills anti-hate and antibias values. For more support, please turn to ADL as a resource. We conduct extensive anti-hate and antibias trainings in schools across America. We put on a range of other relevant events, many through our extensive No Place for Hate program, for improving climates in schools troubled by bias and hate. Our website contains a wealth of material that educators can download for free, including lesson plans, reading lists, helpful guides, and much more on a vast array of topics related to hate, bias, and civics.

## Helping Students Make Sense of Bias, Injustice, and Hate in the News

When incidents of bias, injustice, and hate occur in the wider world, children usually want to be part of the conversation. Although you might wish to protect them and avoid talking about these events in the classroom, these conversations can be productive as long as you conduct them in age-appropriate and helpful ways. If you've worked on creating an antibias learning environment in your school, you're already in a much better position to talk about bias and injustice head-on. Here are some additional strategies to keep in mind:

- *Create a sense of safety:* Develop classroom guidelines that involve the students in the process rather than just handing them a list of rules. The guidelines should address listening, confidentiality, put-downs, and how to deal with bias and stereotyping. Remind students of these guidelines before embarking on a conversation.
- *Express feelings:* Allow students an opportunity to get their emotions out. Help them understand that students with whom they disagree can express their feelings too.
- *Generate questions:* Determine what students know about an event and brainstorm questions to ask with them as well as avenues for answering them. Explore the art of asking questions as a way to develop critical-thinking skills.
- *Share facts:* Clarify misunderstandings that students might have about the episode in question. Help them build tools and skills required to determine what actually happened. Make sure they understand key bias-related terms such as *prejudice, discrimination, equality, equity, target,* and *aggressor.*
- *Learn more:* Encourage students to perform further research into the current event as well as the relevant historical context. For instance, if you're discussing the hatred directed toward the AAPI community, have them research the history of the Asian experience in America, such as the internment of Japanese-Americans during World War II.
- *Complicate thinking:* As important as it is for students to express the strong feelings they have about topics in the news, their learning will benefit if you can complicate their thinking. Expose them to readings that challenge them to hone and refine their points of view. Help them practice building their

case by assigning them to write argumentative essays, letters, and opinion pieces or participate in debates with their peers.

- *Take action:* The news can often feel overwhelming to students. Help them feel empowered by prompting them to take some sort of action, however small. They might teach others, organize a school forum, write letters to politicians, create social media campaigns, raise money for a cause, and much more.

## Making a Difference

Rachel, the twelve-year-old student who was harassed and bullied on the school bus, and her father, Gary, did finally manage to get an official at the county level to take action. Although Rachel's bully never suffered any consequences, the school district brought in ADL some months later to conduct anti-hate presentations in front of thousands of students. Gary attended the presentations, sitting in the rear of the meeting space and watching how the kids were reacting to it. "It seemed like most of them didn't know what the Holocaust was," he says, "judging from a show of hands. But they were definitely engaged, and it was just great."

Looking back on it two years later, Rachel feels that the trainings and subsequent anti-hate efforts at the school (including smaller trainings that she helped lead) made a noticeable difference. Kids seemed to be a bit nicer, and teachers seemed to be taking slurs and other inappropriate behavior more seriously. But Gary isn't so sure. Other antisemitic incidents had occurred at the school since that incident on the bus. A student who sat next to Rachel gave a Heil Hitler salute in view of other students. A teacher had students echo slanderous statements about Jews while participating in an exercise that simulated disagreements between Israelis and Palestinians. Even as ADL training at the school continues, it seems to Gary that many teachers and staff are uninterested and resistant to changing their attitudes. The school clearly has a deeper problem with hate. Gary wonders if educators are all that determined to address it.

Whether or not the other kids and the teachers have changed, there's one person at the school who certainly has: Rachel. Although being bullied and exposed to antisemitism was harrowing, she now feels stronger and more empowered because of her ability to influence the environment at her school. "I had no idea people in middle school thought things like that about Jews," she says. "But I was able to make a change. I see a difference because I didn't sweep it under the rug. And what I say now is don't be a bystander. Don't sit and watch someone get threatened."

I think Rachel's experience leaves us with a mixed but ultimately hopeful picture of American education. Hate is very much a part of growing up today. It's harsh and sometimes violent, and it hurts. But the presence of hate in our schools and in our homes is hardly inevitable. An opening exists to shape the next generation to reject hate and embrace inclusivity and difference. If we work at it as parents and educators, if we take care to model respect for others and thoughtfully instill this value in our homes and classrooms, we can raise anti-hate kids who not only treat others well but stand up when they see injustice happening.

As I remarked earlier in this book, I don't believe that the arc of the moral universe inexorably bends toward justice. We need to reach up, grab it with our bare hands, and bend it ourselves to create the change that we want, including movement toward a post-hate society. A big part of this is educating our kids in how to behave toward others. Given the role of media and other influences, we can't passively assume that kids will naturally glean positive values. We must teach them to become the thoughtful, empowered citizens we know they can be.

As unacceptable as acts of hate are, they can often galvanize kids, parents, and educators to step up and make a difference. But it's unconscionable to wait for hate to rear its ugly head. We can and should *take action now*. If you're a parent, protect your kids by informing them about bias and hate. If you're a teacher, protect the students in your classroom and enhance their education by cultivating a welcoming, inclusive community of learners.

Of course, parents and teachers need not bear the entire burden of turning the next generation against hate. Faith leaders at all levels hold enormous sway over the moral development of both kids and their parents. Some might even say that the creation of a post-hate society hinges on the presence of priests, reverends, rabbis, imams, and other religious figures who preach tolerance over divisiveness, who courageously call out hate when it happens, and who serve as allies for afflicted minorities. These leaders are out there if you know where to look. We will need many more of them speaking far more loudly if we are to keep society from climbing farther up the Pyramid of Hate.

# 10

# FAITH AGAINST HATE

On October 27, 2018, two Muslim-American groups mounted a fundraising campaign on an online platform to help victims of a horrible terrorist attack that had taken place on American soil that very same day. Within six hours, these groups had raised $25,000. Within twenty-four hours, $50,000 had come in. Two days after the campaign started, givers had pledged $150,000.

In the end, over 5,800 people donated more than $230,000 to help victims of the attack pay their bills in the short term. It was a stunning outpouring of love and concern, but what made it especially remarkable was the nature of this campaign and its intended beneficiaries. The victims of this atrocity weren't fellow Muslims. They were Jewish victims of a white supremacist's attack on Pittsburgh's Tree of Life synagogue. Moved by a desire to "respond to evil with good," the campaign organizers sought to reach across the barrier of religion, culture, and perhaps even politics to lend a helping hand at a critical time. Indeed, they stipulated that any excess money raised would go to support interfaith efforts linking the Muslim and Jewish communities.

This campaign was one of numerous interfaith gestures triggered by the Tree of Life massacre. Religious leaders spoke out. The Muslim group Emgage proclaimed its sadness and its support for the Jewish community, while Pope Francis called it an "inhuman act of violence."

Interfaith rallies and vigils took place nationwide and beyond, with the ADL helping to organize a number of them. Pittsburgh's Christian and Muslim leaders loudly condemned the attack and expressed solidarity with the victims, and thousands of local citizens participated in interfaith vigils. Buddhist monks visited the synagogue to pay homage to the victims, as did a group of African immigrants.

Rabbi Hazzan Jeffrey Myers of the Tree of Life synagogue expresses amazement at the tremendous support the Jewish community received from the Pittsburgh community and its diverse faith groups. As he recalls, within hours of the attack, offers of help poured in from all quarters, including mosques, Roman Catholic churches, other synagogues, and Hindu temples. Concerned neighbors asked if Myers's congregation needed money, a place to meet—anything he could think of.

"To me, that's the most powerful story," Myers says. "Every faith community in Pittsburgh reached out. Every single one. I've engaged in interfaith relations my entire professional career. It takes time to build these relationships. They don't happen overnight. But there it was, all of these doors opened instantly, and I'm still trying to enter these doors and build these relationships because there were just so many."

When acts of hate-fueled violence occur, demonstrations of allyship from diverse faith communities can be immeasurably meaningful. The sight of people of different religious beliefs and backgrounds expressing solidarity with victims reaffirms community norms of decency and humanity. But if society is to make inroads against hate, religious groups can't simply rally together once a hate crime has occurred. We must establish allyship as a regular practice, extending hands of friendship to communities outside the bounds of our particular spiritual tribe and speaking out on behalf of inclusion, tolerance, and love. As my friend Wade Henderson, the longtime CEO of the Leadership Conference on Civil and Human Rights, often says, "In order to have a friend, you have to be a friend."

I will be the first to acknowledge that religion has a spotty track record on hate. The Catholic Church erected the architecture of antisemitism in Europe, contriving classic lies such as the blood libel. Despite his efforts to reform the Catholic Church and create a more modern alternative, Martin Luther perpetuated and arguably perfected its virulent antisemitism. More recently, extremist Muslim clerics have fomented radical jihadi ideology. The Islamic Republic of Iran seems to specialize in the most virulent strains of antisemitism as well as vicious hatred of Baha'is, Sufis, Dervishes, and other religious minorities. With Iran's oil production lagging due to domestic incompetence and foreign sanctions, rabid intolerance has become the country's most successful export of the past forty years.

Jewish religious leaders have not been innocent of spreading hate. Extremist Meir Kahane fused anti-Arab racism with violent fascism, radicalizing young American Jews during the 1970s and influencing ugly, hateful political actors in Israel today. One of his crazed followers, Baruch Goldstein, mercilessly gunned down twenty-nine innocent Muslim worshippers in Hebron in 1994. In 2016, the ADL called out the Sephardic chief rabbi of Israel for claiming that "non-Jews shouldn't live in the land of Israel" and that their only purpose was to act as servants to Jews. We demanded that the rabbi retract his remarks and apologize, calling his views "shocking and unacceptable"—which they were.

At the same time, religious leaders from all major faiths have been at the forefront of social movements pushing for inclusivity and tolerance. In the United States, nineteenth-century abolitionists were steeped in biblical teachings. During the twentieth century, Reverend Martin Luther King Jr. called on Baptist leaders across the South to rise up against racism. Numerous Catholic priests and Jewish rabbis marched arm in arm with Dr. King throughout the civil rights movement, driven to do so by their faith.

Today, Jewish rabbis like Angela Buchdahl and Jonah Pesner act as prophetic leaders, organizing and teaching regularly on issues of fairness and justice for all. Christian clerics such as Reverend William Barber and Pastor Rick Warren continue to advocate for justice for oppressed groups and push for an end to hatred. We see the same from prominent members of the Muslim community. As the Muslim academic Dr. Mehnaz Afridi remarked following the 2016 election, "If we don't speak up for one another in these times, then we have failed our own faiths and communities. We need to keep building, organizing and creating positive encounters, and share stories of cooperation that will overcome the bigotry that we as Americans must not tolerate."

Although participation in organized religion has plummeted in the United States in recent years, local and national religious leaders retain great moral authority. Faith communities remain centering influences in society, helping us see past our inflamed emotions and reminding us of our better natures and enduring values. But they need to speak out more loudly and often than they have to date and take more steps to build relationships across doctrinal lines.

If you lead or participate in a religious community, I urge you to rally your fellow believers to take a stand on behalf of inclusivity and tolerance and against hate. Model these values by reaching out to another faith group in a spirit of curiosity and learning. Go beyond differences to explore your common bonds. You might fear that engaging with the Other will dilute your own religious or spiritual identity, but the opposite is true. By coming together to affirm, celebrate, and protect our shared humanity, people of all faiths can realize the highest values of their traditions, which almost universally celebrate pro-social virtues such as charity, forgiveness, empathy, and neighborliness. Fostering civility and peaceful dialogue will allow your faith community to create a safer environment for *everyone* to worship. It will serve as a model for others to follow, helping our society to stop its progression up the pyramid.

## Ripple Effects

Although it is a Jewish organization, ADL has long pursued interfaith relationships. This is rooted in part in one of our core organizational tenets: If diverse religious groups can understand one another, they can reduce interfaith tensions and serve as mutual sources of support. The ADL worked with the Catholic Church during the 1960s to help it construct language that officially repudiated antisemitism. We continue our work today with the Catholic Church, collaborating on a program that teaches parochial-school educators about the Holocaust. We also hold events such as interfaith seders and dialogues. And we participate in the Interfaith Coalition on Mosques, an effort by non-Muslim groups to help our fellow Muslims combat discrimination and hostility from communities who seek to prevent them from building mosques.

If you haven't experienced the personal satisfaction and growth that comes with interfaith discussions, I invite you to try it. I think you'll find you'll become not only more empathetic and compassionate, but also more willing to leap into action as an ally when acts of hate occur.

Some years ago, our Connecticut office organized a series of interactions between Muslim and Jewish women. Although Muslims and Jews often disagree about political issues involving Israel, these women wanted to learn about each other's identities and experiences. Engaging in empathetic, nonjudgmental ways, the women held formal discussions as well as "outreach gatherings," such as attending a Shabbat service or hearing about an individual's pilgrimage to Mecca.

As the women got to know more about each other and their faith experiences, friendships took root and greater understanding flourished. As one of the Muslim women said, "The support and understanding of the group has touched other parts of my life. I feel we are building bridges through compassionate understanding. This expands our network to places other than work and home. We are touching on other beliefs than our own only to realize how great the common ground is."

Some of the group members found themselves more sensitive to acts of hate directed at the Other and more willing to render assistance. One of the Jewish members related witnessing a teenager harassing a Muslim woman in a department store, mocking her hijab. Without thinking, the Jewish woman approached the Muslim woman, took her hand, and led her away from the aggressor.

Such acts of generosity across religious divides — the "ripple effects" of interfaith dialogue, as one of my ADL colleagues observed — don't immediately change our cultural environment. But over time, they do add up. Level 2 of the Pyramid of Hate includes dehumanization of others, and interfaith interactions are an antidote, serving to reveal the other's humanity to us. They foster love, not hate.

## Staying True to Ourselves

You might read my exhortation to interfaith engagement as a call for religious leaders to look away from teaching their own faith and focus instead on activism and mobilization. That is not what I mean to suggest. In fact, countering hate inside religious communities begins precisely by engaging at the level of theology, drawing on your own religious tradition to defeat the arguments of extremist voices.

My friend Abdullah Antepli, a Turkish-born imam who is a professor at Duke University's divinity school and its school of public policy, relates that he often will respond to antisemitic rhetoric in Islam by providing context and texture that neutralizes the hate. As an example, he notes that extremists sometimes arouse hatred toward Jews by citing a story in the Islamic tradition in which a Jewish tribe betrayed the

Prophet Muhammad while he was fighting against Arab nonbelievers. Imam Antepli responds by saying that in this story, eight other Jewish tribes showed loyalty to the Prophet Muhammad and his cause.

Imam Antepli also produces any number of other "inconvenient facts" that complicate Muslim extremists' overly simplistic, extremist view of Islam — the fact that Muhammad attended Jewish religious services, that he had Jewish in-laws, that he ate kosher Sabbath meals, and so on. "What I do," he says, "is turn extremist theology upside down by providing enough context, by knowing the religion as an imam and scholar, and by using authentic Muslim theological truths."

He notes that the simple act of educating people in the nuances and complexities of their own tradition is often enough to turn even hardened extremists away from hate. He tells of one strongly antisemitic religious student he encountered who completely renounced his hatred after a single conversation in which Imam Antepli broadened his view of his own religion. Realizing that his hatred amounted to a violation of Islam's core ethics, this student committed himself to a lifelong process of forgiveness and atonement even though he remained critical of Israel in political terms.

Antepli was especially well placed to help this student and other Muslims who held strong anti-Jewish views because his own story followed a similar trajectory of escaping extremism. Antepli grew up poor during the 1970s and 1980s in the slums of southeastern Turkey, near the border with Syria. As he recalls, dire poverty was extremely demoralizing. It was hard enough to contemplate the misery that awaited him each day and that was palpable all around him, but what made it even more difficult was the knowledge that his forebears had created the Ottoman Empire, a global civilization with a glorious past. The jarring contrast between past and present made him feel like a "perpetual loser" and left him struggling with a burning question: How could he explain the suffering and decay all around him?

Like Damien Patton in chapter 4, Antepli found easy answers and new meaning in vile antisemitism, the poisonous conspiracy that so often animates the beating heart of extremism. When Antepli was a teenager, his father and a teacher in the local public school both urged him to read a children's version of *The Protocols of the Elders of Zion*. Antepli was transfixed by the text, which took the complexity of the world and radically simplified it. Muslims were downtrodden, but he was comforted by the notion that it was all the fault of those evil, scheming, pernicious Jews who, he now believed, were bent on world domination. It was a seductive form of hate, as dangerous as what Damien had been taught by his right-wing fellow travelers.

Antepli devoured other virulently antisemitic screeds, such as Henry Ford's *The International Jew* and Hitler's *Mein Kampf*, and quickly developed a deep, even murderous antipathy toward Jews, Israel, and the West (which in his mind was under Jewish influence). Having decided the Jews were irredeemably evil and worthy of destruction, he yearned for their annihilation and the destruction of the State of Israel. In fact, during the Palestinian intifada of the late 1980s, he burned Israeli flags and bought his friends ice cream each time he learned that a terrorist had murdered an Israeli civilian or soldier.

Antepli might have remained mired in hate had he not developed a new interest in Islam around this time. Like others of his generation, he became deeply religious, a break with the secular traditions of his parents. As a college student, he learned Arabic, studied the Koran and other texts, learned Islamic prayer rituals, and immersed himself in Islamic history. But the more he learned, the more he struggled with an internal conflict: Islamic teachings about love, compassion, forgiveness, and mercy moved him to his core, but he couldn't square them with the poison of antisemitism that had organized his thinking up to that point.

As Antepli continued his studies, Islam won out over intolerance. Learning about the close relationships between early Muslims and Jews, he came to realize that he'd been manipulated by antisemitic texts and that their simplistic narratives were fundamentally wrong and misguided. He studied Jewish texts and perceived that they, too, contained a divine spirit. "Good religious education slowed me down on the highway of hate," he says.

By the time he graduated from college, Antepli had put his life on a different path. He had undergone "the mother of all changes" and become a "recovering antisemite." The process of recovery continues to define and give meaning to his life in much the same way that rehabilitated white supremacists find meaning and derive a kind of forgiveness through interfaith work.

Antepli notes that when you encounter hatred so early in life, you can push it back, but you can never eradicate it fully from your psyche. You must constantly fight to prevent the virus of hate from sickening you again. Over the decades, he's succeeded in keeping hate at bay by continuing to educate himself and by interacting with God-fearing, moral Jews, especially religious ones. These interactions have left him feeling sad and ashamed about his past beliefs but also grateful that he has been able to free himself to a large extent from hatred and see through the lies he was fed, lies not just about Jews but also about members of the LGBTQ community, Armenians, and other minorities. Recognizing the power of interfaith engagement to counteract hate, he has dedicated his career to fostering connections between Jews and Muslims, most notably as a fellow and codirector of the Muslim Leadership Initiative at the Shalom Hartman Institute in Jerusalem.

Antepli's story teaches us that in a moment of political polarization, extremists come in many forms and will use any set of beliefs, including faith, to rationalize their irrational intolerance. And yet, like Damien's escape from white nationalism, Antepli's disavowal of Islamic extremism serves as an inspiration, reminding us what we can accomplish when we are willing to look inside our souls, become aware of our biases and hatreds, and work to overcome them. Leaders from all religious faiths must recognize the extremist voices in their midst and develop their own narratives of tolerance, nuance, and complexity, grounding them even more firmly in scripture and ritual. Looking beyond theology, I believe that religious leaders of all faiths should embrace more activism, but as an addition to their immersion in religious doctrine and ritual rather than an alternative to it. Activism and adherence to tradition need not be mutually exclusive. On the contrary, activism can be a way to bring religious dogma alive.

One of my favorite anecdotes concerns Rabbi Abraham Joshua Heschel and his memorable statement upon returning from the Selma-to-Montgomery march beside Dr. King: "I felt my legs were praying." That had historical precedent: Frederick Douglass likewise is said to have remarked, "Praying for freedom never did me any good 'til I started praying with my feet." We need to stand up and fight, not just pray or meditate in a disinterested way. Society would benefit if all of us, in addition to our usual forms of observance, took more time to pray with our feet on behalf of justice and inclusion.

I would note that it's in the interests of faith communities themselves to encourage praying with one's feet. Experts attribute declining rates of attendance among the younger generation to their preference for activism and social justice over ritual and rote observance. To the extent that religious leaders can find ways to connect the principles that appear in holy scripture to the realities of the world today, they can reclaim that sense of relevance. Again, that doesn't mean a community should forgo ritual. Contemporary political values are a poor replacement for timeless traditions of faith. But there may be no purer or more powerful way to forge connections between scripture and the present day than by speaking out on behalf of members of *other* faith groups.

Most religious communities won't change overnight and embrace an agenda of transformation. Communities must work to find a middle ground, doing what they can while still staying true to their beliefs. As I've seen firsthand, meaningful change can take place in this middle ground, and even groups with very conservative beliefs can take stands in support of tolerance and unity.

When I came on board at ADL in 2015, I learned that forty-five states had hate-crime laws on the books but that Utah had not passed such legislation. Some attributed this to concerns among Utah legislators that such a move would indicate tacit approval of the LGBTQ community. The legislators seemed reluctant to proffer such approval, given that the Church of Jesus Christ of Latter-Day Saints, the largest faith group in Utah and the state's preeminent political force, ardently disapproved of homosexuality. Still, I held out hope that change was possible, and in 2019 I jumped at the opportunity to spur it along.

At the invitation of church elders, I flew to Salt Lake City and spent several days with them. I learned a great deal about the church, but I also took the opportunity to advocate for the ADL's model anti-hate legislation. I spoke with elders at meals and meetings, gave a well-publicized speech at a legal society attended by a number of state legislators who also were members of the church, visited charitable projects that served the entire community, and spoke at Brigham Young University in front of hundreds of students, almost all of whom were Latter-Day Saints.

On these occasions, I talked about the New Testament and Jesus's exhortation that we love our neighbors. I discussed how hate crimes affect the less fortunate and disempowered, appealing to my listeners' innate sense of justice. I noted that in the absence of state anti-hate laws, the federal government was more likely to intervene when crimes were committed — an act perceived as overreach in this deeply conservative state.

The trip wrapped up, and as I headed to the airport, my phone buzzed with some surprising news: a local paper was reporting that the

church had made a 180-degree turn and intended to put its weight behind anti-hate legislation. Seeing that text message was one of the most
rewarding moments I've had in my work at the ADL. The chance to exchange views and empathize with people who had profoundly different
ideas and find common ground—it was really amazing. To this day, I
keep a copy of the Book of Mormon in my office. It was given to me by
church elders, and I cherish it as a sign of both their goodwill and the
wider possibilities of dialogue and engagement.

To be clear, I'm not suggesting that in this one trip I single-handedly
convinced the Church of Jesus Christ of Latter-Day Saints to clamp
down on hate. The elders had already been on the cusp of shifting their
views on the issue. It might surprise some critics, but the church is actually an extremely diverse and dynamic organization, accepting of people of all ethnicities, a reality I observed while visiting a church-funded
food pantry in Salt Lake City staffed by refugees. Gazing at the vast
number of flags hanging on the walls, each signifying the home country of one of the food pantry's workers, I felt like I was standing in the
lobby of the United Nations.

When it came time to protect people, including members of the
LGBTQ community, from hate crimes, I happened to be the right person at the right time to give the church the final nudge it needed. But
that's my point—people in your religious community might not be as
locked into their beliefs as you think. You might be in the perfect position to help them reach new levels of openness, toleration, and respect
and to make a difference in the wider world.

## In Search of Great Leaders

More generally, hope, renewal, and progress remain possible at all times,
including when the unthinkable happens. As painful and horrible as it
is, mass killings and other hate-fueled violence can leave in their wake

new bonds of fellowship as well as a new determination to offer solace and support and fight back.

Reflecting on the aftermath of the attack on his synagogue, Rabbi Myers noted a positive development — the links that the Tree of Life synagogue had forged with communities of other faiths that had also suffered from the epidemic of hate. He himself had developed a friendship with Pastor Eric Manning of the Mother Emanuel AME church in Charleston, South Carolina, where in June 2015, a white supremacist ruthlessly gunned down nine Black victims assembled for a Bible study class. Synagogue members also connected with victims of the 2016 Pulse nightclub shooting in Orlando, Florida, the deadliest anti-LGBTQ crime in American history, and with Sikhs in Oak Tree, Wisconsin, who were targeted in a 2012 massacre that killed six innocent people at the gurdwara where they worshipped.

Remember the successful fundraising effort mounted by the Muslim community in the wake of the Tree of Life attack? Following the 2019 attack on two mosques in Christchurch, New Zealand, the Tree of Life synagogue returned the favor, holding a GoFundMe campaign that raised almost $65,000 for victims. "We feel compelled to come to the aid of those communities," the synagogue's congregants explained, "just as our Jewish community was so compassionately supported only a few short months ago by people around the world of many faiths." Members of the congregation also visited a local Islamic center and participated in classes to show their support.

As Rabbi Myers relates, the tragedy at his synagogue not only sensitized his congregation and himself to the suffering of others but awakened him to a new mission in life: combating hate. Prior to the attack, he struggled to determine what his responsibility was as a faith leader in terms of hate crimes. Was he supposed to speak out about guns? Advocate for nonviolence? After the attack, he received what he sees as a "divine message" regarding his purpose. "God said to me, 'Jeffrey, it's not your turn yet. I don't want you yet. I have stuff I need you to do.'"

As Rabbi Myers realized, the problem facing our country and indeed the world wasn't just the accessibility of firearms or even widespread racist beliefs. These were all symptoms of a larger problem — hate, or as Myers prefers to call it, the "H-word." "That's the root of it," he says. "If you think about a weed, you have to carefully cut around it and excise it, otherwise it's going to continue to grow." To stop acts of hate, we ultimately need to prevent the spread of hateful words and ideas, because left unchecked, those words lead to actions. Myers's mission is to fight back against hate.

And it isn't just about asking people to tone down offensive language. Rabbi Myers explains that it's about "demanding that those in positions of authority restore civil discourse." It's also about educating ourselves about one another and, in particular, opening ourselves up to different faith traditions. "We need to recognize that no matter what our faith is, that's only part of the planet. If you're a Roman Catholic, there's still the rest of the world. What do you know about their faith, their observances, their traditions? I think if we get to know our neighbors better, that will ease away the mistrust, the fear, the loathing of things we don't understand. And that in turn leads to respect for others."

For Rabbi Myers, respect is key. It's no longer enough for us to tolerate one another. We need to show mutual respect. And we can do that only if we take the time to understand one another. This logic leads to a sense of urgency on Rabbi Myers's part, a belief that people can't just sit back passively; they have to be actively engaged to reverse the dangerous forces unleashed in society.

We also need forceful leadership. Where, he asks, are the people today who can speak with the moral clarity and persuasiveness of a Dr. Martin Luther King Jr.? "We don't have such people," he says. "I don't see them. It takes immense perseverance and strength to just go on and make that your life's mission because you know you have to."

Rabbi Myers is hardly alone in this belief. Imam Antepli bemoans "an incredible lack of moral courage and prophetic voices in the Jewish and Muslim communities" when it comes to countering hate. Leaders are quick to call out bigotry expressed by members of other faith tradition, but often fail to take a strong stand when their coreligionists spew hateful beliefs. "I think part of religious leadership is to be a prophet, in the sense that your mission and responsibility is to make people uncomfortable, to put a mirror to their ethical and moral failings."

Many of us would not define a prophet along these lines, but if you're a faith leader, I challenge you to try your best to become a truly prophetic leader. Show the courage and vision of a figure like Moses, who brought his people out of bondage and back to their homeland. Speak up on behalf of love and inclusivity more loudly and intensely than ever before. Set an example by examining your own biases and working to overcome them. Lead your community to show greater openness and neighborliness toward people of faiths different than yours. Challenge your community to make itself vulnerable and learn from others. Call out hatred when it becomes present among your flock.

Working together, we can connect with people of different faiths, build strong relationships with them, and learn to respect them. That in turn will allow us to keep society near the bottom of the Pyramid of Hate and out of grave danger. "The H-word is the biggest problem in the United States," Rabbi Myers says, but it's also a problem that we can solve. "We're capable of so much, we really are, I believe that."

And so do I.

I hope and pray that our faith leaders will find the strength and determination they need to push back against hate and replace it with love. I hope and pray that our business leaders will as well. Like faith leaders, titans of industry possess moral authority and a centering influence, one that they have too seldom wielded in the fight against hate. Further, they possess immense resources and persuasive power through their companies and brands, not to mention their entrepreneurial energy and capacity for invention. To ensure that the worst of human experience doesn't happen here, the business community must stand up to bigotry and intolerance as never before. They must make it clear for all to hear: Hate has no place in our society, including in our workplaces, our organizations, and our public discourse.

# 11

# BUILDING BETTER BUSINESSES

In March 2021, the Republican-controlled legislature of Georgia passed a new law that indisputably restricted the ability of citizens in the state to vote. Voters now have less time to request absentee ballots. State officials can assume control over local electoral bodies more easily. And you can't even approach a voter waiting in Georgia's long lines at polling places to offer them food and water, a tactic credited with increasing turnout in the 2020 presidential elections.

Many voting-rights activists vigorously opposed the law, one of hundreds proposed in 2021 in GOP-controlled statehouses that, under the guise of preventing fraud, would limit the ability to vote. Critics called out these laws as naked attempts to disenfranchise Blacks, Latinos, and other groups who had voted in large numbers against President Trump in the 2020 cycle. One state senator proclaimed the Georgia law "the Christmas tree of goodies for voter suppression." Newly elected President Biden called Republican efforts "the most pernicious thing" and opined, "This makes Jim Crow look like Jim Eagle."

Prior to the Georgia voter-suppression law's passage, the ADL spoke out against the legislative bills that gave rise to it, observing that they would "undoubtedly disenfranchise thousands of Georgians with a disproportionate impact on racial minority, low-income, elderly, rural, disabled, and student voters." As we further noted, the very premise of these bills—that widespread fraud had taken place in 2020 and that the Democrats had "stolen" the election from Donald Trump—was a complete fiction, the stuff of vile conspiracy theories and disinformation.

Advocating on behalf of voter rights is not new for the ADL. Mindful of the shameful efforts made by racist legislatures to disenfranchise Blacks and other minorities during the nineteenth and twentieth centuries, we view support for voting rights as a critical element in the fight against hate and bigotry. ADL staff and supporters mobilized during the U.S. civil rights movement in the 1950s and 1960s. In 1965, ADL leaders joined the Reverend Martin Luther King Jr. in his famed march from Selma to Montgomery on behalf of voting rights, advocating later that year for the Voting Rights Act (VRA), which put an end to many previous voter-suppression measures. Since that time, we've supported extensions of the VRA and spoken out against the new voter-suppression measures that arose after the 2013 Supreme Court decision *Shelby County v. Holder* negated key provisions of the VRA. In this effort, we've worked as part of coalitions of civil rights groups, religious groups, students, and concerned citizens.

But there's one group that traditionally hasn't joined these efforts: businesses. With some exceptions, much of corporate America took a pass on civic activism during the civil rights era. And it continued to stay quiet for decades on issues related to racism, discrimination, hate, and extremism.

This hesitation began to change during the Trump era, and the change intensified noticeably in 2021. The sudden willingness of the business community to speak out publicly and forcefully against voter suppression in Georgia and other states stands as one of the most striking developments of this latest chapter in the fight for voting rights.

When Georgia's bill was first passed, Delta, Coca-Cola, and other large corporations headquartered in the state stayed noticeably silent.

This response—or lack thereof—raised the ire of activists, who noted these companies' loud embrace of racial justice during the recent Black Lives Matter protests. Then two prominent Black executives, Merck CEO Kenneth Frazier and former American Express CEO Kenneth Chenault, organized dozens of their peers to sign a public statement urging corporate America to "take a nonpartisan stand for equality and democracy" and push back against restrictive voting laws.

Their statement seemed to break an invisible logjam in corporate America. Within days, hundreds of companies came out publicly against restrictive voting laws, including big names like Amazon, Starbucks, and Microsoft. Major League Baseball decided to relocate its 2021 All-Star game out of Georgia to protest the state's new law. Other organizations took similar actions.

Republican politicians pushed back hard. Florida governor Ron DeSantis accused companies of "getting in bed with the Left, the corporate media and big tech." Senate majority leader Mitch McConnell told corporate America to "stay out of politics." Donald Trump released a statement urging his supporters to boycott MLB as well as "all of the woke companies that are interfering with Free and Fair Elections."

Although the ex-president and his supporters might regard public gestures of support for voting rights as political acts, I see them as important statements of principle. At issue isn't so much how you vote but what you value. The ability to participate in our democracy as a voter remains the most fundamental right in our constitutional system, the lever that makes our democracy work. Restrictive voting laws might serve this time to disenfranchise Democrat-leaning voters, but legislatures could one day apply the same tactics to prevent Republican-leaning citizens from voting. Regardless of party, we should reject restrictive voting laws as not only unjust but dangerous, a pathway to tyranny.

In the case of Georgia's voter-suppression controversy, the business community deserves applause for doing its part to push back on systemic racism—at the ballot box and within the ranks of corporate America itself. We should credit business leaders for the stands they've taken on other issues related to equality and civil rights, such as the decision of many businesses to boycott North Carolina in 2016 over its restrictive transgender bathroom law and that of the more than twelve hundred advertisers who joined our Stop Hate for Profit campaign in 2020.

And yet, as important as these gestures have been, they aren't nearly enough. Companies large and small and their leaders must do much more in the years ahead to fight intolerance, bigotry, and violent extremism. The business community must show a new kind of moral leadership, standing up and affirming to stakeholders that there's truly no place for hate in this country.

Business leaders must understand their position in society and the power they possess, as well as the responsibility that comes with it. At a time when Americans are polarized, when our political system has become mired in dysfunction, and when trust in most other social institutions has tanked, business still retains authority and influence. The 2021 Edelman Trust Barometer, a survey of tens of thousands of people in more than twenty-four countries, found that people trust business more than they do governments, media, or NGOs.

We can lament this reality and work to reestablish trust in the latter three sectors of society, but for purposes of fighting hate we must recognize the influence that companies and brands enjoy. The business community must mobilize its moral and cultural capital to help prevent society from moving up the pyramid. My friend Richard Edelman, who conceptualized the trust barometer, puts it well: "Business has to help the country knit itself back together, to be the connective tissue, while the politics are sorted out."

## Building a Case for Socially Responsible Business

I've presented a moral argument for why business should take a stand on hate, but if you run a business, I'd also like to appeal to your self-interest. In recent years, social expectations of businesses have changed dramatically. Whereas stakeholders used to find it acceptable for a business to focus purely on its products and services and let the government confront challenges like hate and intolerance, that's no longer the case. Customers, employees, and others increasingly reward businesses that deliver on social issues and penalize those that fail to do so.

That business hasn't done more to engage on social issues might be due to a particular way of thinking about capitalism that prevailed during the late twentieth century. In 1970, University of Chicago economist Milton Friedman argued persuasively that "the social responsibility of business is to increase its profits." Companies, Friedman wrote, weren't meant to save rain forests or take political stands against racial oppression. In a capitalist system, firms served a single purpose: to make money for their owners. Anything else would send society on a fast track to socialism. As Friedman remarked, "The use of the cloak of social responsibility, and the nonsense spoken in its name by influential and prestigious businessmen, does clearly harm the foundations of a free society."

During the 1970s and 1980s, this thesis of the company-as-profit-machine became the dominant ideology of American capitalism, shunting aside earlier traditions that regarded companies and communities as possessing a synergistic social and moral responsibility. During the 1980s, the Reagan administration sought to unshackle business to kick-start an economic recovery. Drawing heavily on Freidman's thesis, it adopted policies designed to enable corporations to focus fully on maximizing profits for shareholders. The administration grafted on the optimistic notion that unbridled economic activity would benefit everyone by allowing wealth to "trickle down" from society's upper echelons.

Unburdened by moral responsibility, CEOs like Jack Welch, Al Dunlop, and others practiced a ruthless style of capitalism. They busted unions, bought and sold companies like trading cards, and ravaged the environment. During the latter decades of the twentieth century and on into the twenty-first, the rich got richer and the poor poorer. Climate change and other environmental issues worsened. Public infrastructure decayed and crumbled. And CEOs, whose compensation was in large part tied to stock performance, did very well for themselves.

There were outliers, of course—pioneering, socially responsible brands like the Body Shop, Patagonia, Stonyfield Farm, and Whole Foods Market that, during the 1980s and 1990s, attracted consumers by doing good *and* doing well. During the early 2000s, other socially responsible companies cropped up, businesses like Honest Tea, KIND Snacks, and the company I cofounded, Ethos Brands. Fueling this newer generation of firms was a sense that climate change was worsening, that globalization wasn't all it was cracked up to be, that the world had an unaddressed poverty problem, and that companies could operate more ethically and sustainably if they tried.

Over the past decade, socially responsible business took on even more prominence as technologies like solar power, wind power, and electric cars became attractive commercial opportunities. The sharing economy emerged as a way to make more efficient use of existing real estate (as in the case of Airbnb) and cars (Uber and Lyft). Organizations began to create measures and standards for sustainable business, and the third-party B Corporation designation made businesses legally accountable for serving society as well as shareholders. Newer companies advocating social responsibility arose, including firms like Warby Parker and Toms Shoes that, echoing our efforts at Ethos, championed a one-for-one model linking consumption with social causes (Warby Parker, for instance, gave away a pair of glasses for each one sold. Similarly, Ethos donated a portion of proceeds from every bottle of water it sold to fund clean-water projects in developing countries that suffered from water scarcity).

These firms were still the exception more than the rule—the vast majority of businesses continued to push profit over people. But a turning point came in 2019 with a public statement by one of the country's most influential lobbying groups, the Business Roundtable (BRT). The BRT's members include the CEOs of America's most prominent corporations, a veritable Who's Who of the establishment. So it turned heads when the BRT issued a statement breaking with Friedman's thesis and adopting a notion of capitalism responsive to "all stakeholders." The purpose of a company, the BRT proclaimed, wasn't primarily to earn profits for shareholders. Rather, members of the BRT recognized "a fundamental commitment to all of our stakeholders," including employees, customers, suppliers, and communities.

Social responsibility has moved even farther into the mainstream with the emergence of BlackRock CEO Larry Fink as a powerful voice for change. Charismatic entrepreneurs like Howard Schultz and Marc Benioff were already trumpeting the need for a more enlightened model of business, but Fink came across as more deliberate and measured— the prototypical asset manager. And as the world's largest investment corporation, with more than four trillion dollars under management, BlackRock indeed was a deliberate and measured organization, exercising tremendous influence in markets worldwide.

For years, Wall Street analysts scoured the annual letters Fink issued to his shareholders for insights and tidbits. These letters discussed the need for more long-term thinking, a stance that seemed perfectly suited to BlackRock's self-interest. But in 2019, Fink's letter specifically elevated the issue of social purpose. Going forward, he stated, BlackRock would consider social and environmental impact when making investment decisions. Alongside the BRT proclamation, this second bombshell alerted the world that business as usual would no longer suffice.

It's worth noting that by this time, the bar for what constituted a "socially responsible business" had risen dramatically. It was no longer enough for companies to produce products sustainably using fair-trade labor practices. Nor was it enough for the product itself to confer a social benefit (healthy organic food versus junk food; renewable energy versus fossil fuels). The most socially responsible companies also had to pay attention to the wider social context in which they operated. They needed to care about community and show a willingness to fight for it.

Today, with societies polarized and elected officials failing to offer principled leadership, businesses of any size in any industry can no longer shrink from taking moral stands. Some issues, like hate, extremism, and racial justice, might be fairly easy to address, while others, such as threats to our democratic process, are in trickier terrain. But that does not exempt the private sector from taking a position. Indeed, the status quo of nonengagement is, in fact, a position, one that a growing segment of society no longer finds satisfactory.

If hate overwhelms us and leads to violence and instability or if authoritarianism snuffs out liberal democracy, markets will unravel and your business likely will suffer. Beyond that, we see time and again that the most talented young people today actively seek to work at companies that stand against hate and that treat their people equitably and with respect. Ample data demonstrates that companies perform better when they have diverse workforces who feel well treated, respected, supported, and empowered to contribute. And as my partner Peter and I saw firsthand when growing Ethos, consumers want to spend their money on products from companies that are guided by principles as much as profits. They want to believe in the brands they use and will exhibit extraordinary loyalty toward those that satisfy their basic needs while also serving the public interest.

The data suggests that consumers care specifically about matters of fairness and justice. Polls have shown that large majorities of Americans want CEOs to take moral stands related to racial equality. Research reported by Edelman in May 2021 found that almost 80 percent of those surveyed expected action on the part of CEOs to combat racism. And more than half of those surveyed felt that leaders were responding well by cutting business ties with states that instituted discriminatory laws. The same research found that many customers rewarded brands that acted against racism and penalized those that didn't and that employees were less inclined to work for companies that punted on racial equality.

Social pressure on business to take moral stands on a wide range of issues is mounting. Research published in 2018 revealed that virtually all respondents believed that companies had the power to improve society, and over 70 percent reported an increase in their expectations of business. A 2020 study found that over half of Americans felt that companies should take public stances on issues affecting society.

Leaders themselves understand this pressure, with over 80 percent in one survey agreeing that business must step up and play a key role in addressing social problems. Vast majorities of leaders in this same survey also acknowledged the many business benefits that accrued when standing up on social issues, including more customer loyalty, better employee recruitment, and better financial performance.

In short, a new form of capitalism is evolving before our eyes, one that breaks decisively from the shareholder-maximization approach. A compelling case now exists for running businesses in morally responsible ways and, specifically, for taking public stands against bigotry and intolerance. Fighting hate isn't only the right thing to do. It's vital for the future success of any enterprise.

## What Businesses Can Do

There are many steps companies can take to push back against hate. Most obviously, they can *engage in public advocacy*. Like Ken Frazier and Ken Chenault, CEOs can use their voices to denounce hate. They can pressure others in positions of influence to speak up as well—not to censor speech, but to make clear that hate speech, while legal, is deeply offensive. They can help expose the haters to public scrutiny, knowing that most Americans will reject hatred when they see it.

If you're a business leader, take note. The next time you see a campaign, such as Stop Hate for Profit, that is designed to compel action against intolerance, judge it carefully and consider joining. And if you see companies, governments, or others acting in ways that sanction or support hate but you don't see anybody doing anything about it, start a campaign in the communities where you do business. Remember, you as a leader have an extraordinary amount of influence, not just due to your sway with customers and others in your communities, but because of the power of your purse. Through advertising muscle, large companies can help determine what information networks like Fox News and Facebook will and will not permit on their platforms. They can hold these services accountable as few others in society can because corporate dollars drive their business models. And don't believe that the size of these platforms insulates them. As important as revenue pressures are, the sheer force of reputational pressure matters too, and every advertiser can play a role in bringing that to bear.

If you run a large firm, do your utmost to ensure that hatred and conspiracy theories are not amplified. Choose to pause or even pull your ads, not just from problematic programs, but from entire networks that don't respect all people or that repeat baseless conspiracies that endanger all of us. Use your conversations with leaders of these networks to push them to do more. Use your public platforms to demand the same.

If you run a small local company without a big ad budget, your voice matters too. You can put up a sign in your window indicating that all are welcome. You can support local political candidates who back inclusion and tolerance. You can create or join an anti-hate coalition in your town. A couple of decades ago, sponsoring your local Little League team might have been enough to establish you as a supportive member of the community. Today, you must do more. You have a stake in ensuring a civil and welcoming environment in your town. This isn't necessarily about being "woke" or taking steps that make you feel uncomfortable. Find a way to embrace inclusivity, tolerance, and civility that fits your values and beliefs.

The fight against hate is the fight of our time. My challenge to you is simple: Commit to this fight. Choose a side.

One of the greatest concerns business leaders have when it comes to choosing sides is the fear that they'll alienate certain segments of the public or even infringe on others' First Amendment rights by taking a moral stand on hate. Let's consider the constitutional issue for a moment. I sympathize with concerns about freedom of speech, but remember, we all have the legal right to decide not to sanction or support speech with which we disagree. And we also have a moral obligation to rid our communities of speech that does real harm by validating violent extremism or even helping it to spread.

In October 2020, the *New York Times* published an opinion piece reexamining the role women played during the mid-1990s in the Million Man March. The piece glorified Louis Farrakhan, organizer of the event and a noted misogynist. Shamefully, the opinion piece didn't so much as acknowledge a critically important fact about Farrakhan—his intense, long-standing, and publicly proclaimed loathing of Jews.

Farrakhan isn't just another antisemite. He's arguably been the lead-ing public antisemite in America over the past two decades. Not long ago, he even targeted me in one of his hateful rants, literally calling me Satan and then extending the libel to Jews in general. It was dangerous and defamatory but classic Farrakhan. And yet, numerous celebrities and sycophants have laundered and shared his anti-Jewish invective. Imagine what would happen if the *New York Times* published a piece lauding Harvey Weinstein's role in creating the independent film in-dustry and neglecting to mention his horrific history of sexual assault against women. Imagine if the *Times* published an op-ed celebrating Strom Thurmond as one of the Senate's longest-serving members and failing to inform readers that he was an arch-segregationist and racist. Would the *Times* ever even conceive of publishing such pieces?

After reading the op-ed, I sent a tough note to the *New York Times'* publisher, A. G. Sulzberger, and met with the head of the paper's ed-itorial page. The editor agreed with me and apologized, acknowledg-ing that the *Times* should have declined to publish the piece. The paper would publicly rectify the error, she said. Sadly, it never did. As I later learned, the *Times* actually promoted the editor who had published that piece.

You might claim, as people and companies often do, that the *Times* had every right to publish this piece, as it's protected by the First Amendment. That's certainly true. If you post a message on Facebook giving people a public figure's address and exhorting them to blow up his home, that's harmful speech that the Constitution doesn't protect. But simply praising a notorious bigot in a way that doesn't pose a clear and present danger to anyone's well-being—that's hate speech, but it is indeed protected. In this instance, however, I would contend the *Times* had a moral obligation not to normalize Farrakhan by ignoring his vile ideology. The *Times'* failure to call that out essentially excused it. And

I would contend that all business leaders have a moral obligation to join activists such as myself in pressuring content providers to make editorial decisions that will heal our country, not further divide it.

Protecting our quality of life as well as our democratic values, laws, and institutions requires that we shut down the haters and deny them a voice as much as possible. Although we might never eradicate hate speech, we can make progress by pushing back on prejudiced ideas, not privileging them as if they deserved a prime-time slot. Television programs should choose not to book bigoted speakers. Social media services should opt not to use algorithms that amplify intolerance. Magazines should not publish op-eds by activists who replace facts with fiction and compassion with conspiracy theories. To be clear, this isn't censorship—it's standards. And, again, these people can spread their ideas, just not in the spotlight. Let's have the self-respect to honor our liberties *and* conduct ourselves and our businesses in reasonable, respectful ways.

Beyond taking part in public activism, you can help build a more respectful and tolerant society by infusing an anti-hate agenda into all aspects of your business. It's not enough for companies to join the occasional boycott or invest a few dollars in bias-sensitivity training, as most currently do. Companies should consider taking a number of steps that don't interfere with core operations but enable the culture to embrace inclusivity more effectively. These steps include

- pursuing diversity and inclusion in hiring strategies and encoding it into the process of professional development and promotion
- marketing products and services in ways that respect everyone (for instance, ensuring effective presentation in ads and not advertising on platforms that allow hate speech to spread unchecked)
- embedding equity in their supply-chain strategy by buying from companies that themselves treat their workforces equitably

- making sure that they reinvest in the diverse communities where
  they operate.

Sometimes companies must sacrifice short-term profits in the service
of combating hate. Painful trade-offs are familiar at a purpose- or val-
ues-led business. When I helped lead Ethos Brands, my cofounder and
I decided to sell our business to Starbucks in 2005, right as we were beginning to scale, rather than grow the brand further ourselves and aim
for a larger exit down the road. We might have made far more money
had we held on to the business and sold it once we had achieved more
scale. But our purpose wasn't just to maximize profits; it was also to
help alleviate the world water crisis. We believed that Starbucks was the
rare corporate parent with the resources and values to sustain our brand
and grow it for years to come. We made the decision that best served
our core mission: helping children get clean water.

More companies need to infuse their decision-making with purpose
and, specifically, with a commitment to fight hate. They need on occasion to make decisions that might be tough in the short term but that are
*right*. I'm not asking companies to take steps that threaten their long-
term viability. Delta Air Lines and Coca-Cola won't pull their headquarters from Georgia or stop serving that market as an act of protest
—that's simply not realistic. But they could perhaps quite viably decide
to relocate an intended expansion elsewhere without much long-term
pain. They could relocate a regular conference or defer a planned pro-
motional activity. During the North Carolina transgender-bathroom
controversy, the NBA didn't demand that the Charlotte Hornets relocate from the state, as that would have upended the franchise's business.
But the league did change the site of the All-Star game as an act of
protest.

The shift toward more social responsibility on the part of business is
welcome news. With governments shrinking and institutions receding,

businesses play an increasingly important social role. We should hardly surrender our fate to unelected corporate boards; government and civil society must step forward and do their part. But to ensure their own survival and to support the health of our democracy and civil society, the business community must exercise moral leadership. Some corporate leaders have sought to do so for years around issues like sustainability, but it's time that they devote more of their energy and advocacy to pushing back hard against hate. And when they don't, the rest of us must pressure them into building the decent, respectful society *we* want to see.

And yet, CEOs need to exercise their power wisely. When they opt to engage, they must educate themselves and act judiciously. If they don't, they might overreach in the name of social responsibility, damaging their hard-earned credibility.

Unilever subsidiary Ben and Jerry's found itself in exactly this position in July 2021 following its sudden decision to pull its products out of the West Bank. In most circles, the decision sparked confusion and concern. The brand's activist board chair, who had previously posted anti-Israel sentiments on social media, initially suggested that the business supported the BDS (Boycott, Divestment, Sanctions) movement. Unilever CEO Alan Jope explicitly informed me that the company was not embracing BDS. The brand's founders, Ben Cohen and Jerry Greenfield, tried to find a middle ground, defending the move but distancing themselves from BDS. Nonetheless, as of the writing of this book, Unilever appeared unable or unwilling to reverse the decision. The brand is now linked to an insidious effort to delegitimize the Jewish state, and it faces widespread opposition and the prospect of legal action.

It seemed Ben and Jerry's management and Unilever corporate leadership had failed to educate themselves and consider the complicated issues before acting. Many observers who strongly support a two-state solution and want to stop the expansion of Israeli settlements reasonably argue that, on the heels of a violent surge of antisemitism in America triggered by extremists who used the conflict as cover, the brand's action was not well timed, let alone morally defensible. I've always been a big fan of Cherry Garcia, but I concur. The brand's decision is ugly, especially when Ben and Jerry's continues to sell its products in countries with arguably more egregious records of human rights violations, like Brazil and the Philippines.

Beyond educating themselves about the issue and consulting a wide range of stakeholders before acting, companies would do well to implement initiatives consistent with their core competencies and operational design. Leaders can't simply bolt a purpose onto a business as an afterthought. It is most effective when it flows logically and understandably from the enterprise's core architecture and long-standing practices. In the end, Ben and Jerry's failed to engage creatively on the Israeli-Palestinian conflict in a manner on brand for its entrepreneurial and imaginative business.

If its leaders sought to make a statement, they might have brought people along more effectively and ultimately achieved far more impact by using their scoop shops to host thoughtful dialogues about the conflict; they could have given grants to organizations promoting peaceful coexistence in Israel or launched a creative new flavor whose profits supported such organizations. Instead, they made a poorly considered, politicized decision that radiated hastiness and misunderstanding. It is a short hop from social responsibility to serious irresponsibility if companies fail to use their power wisely.

## Businesses Stepping Up

In the spring of 2020, weeks or months before the media began to take notice, we at the ADL became concerned about a rise in hate directed against Asian-Americans. We started to pick up more references online to the "Wuhan flu" and the "Chinese flu," and our field offices began to send in reports about harassment of local Asian-Americans. I reached out to some established Asian-American and Pacific Islander activist groups and expressed a willingness to help.

These groups declined the offer for various reasons, but about this time, Daniel Lubetzky, an ADL board member, contacted me and said that he had a friend who knew some prominent Asian-Americans who felt very concerned about hate and wanted to know if the ADL could help. In May, I delivered a presentation via Zoom to about two dozen Asian-American business leaders. I shared my fear that the hate directed at the AAPI community would worsen because anti-Chinese sentiment seemed poised to increase due to the COVID-19 pandemic and rising tensions with Beijing ahead of the 2020 election. I implored them to counter this potentially lethal trend by organizing themselves, and I shared highlights from the ADL's history of battling discrimination and hate over the decades. In so many words, I suggested that they create their own version of the ADL.

Afterward, I received a call from Li Lu, one of these leaders, who also had been among the original organizers of the 1989 Tiananmen Square protests. After being smuggled out of China, he landed in New York, eventually graduated from Columbia University, and founded a highly successful investment firm. Li Lu asked if I would help him and some of his peers set up an Asian-American ADL, assist in hiring and training staff, and share our technological know-how to track Asian-American hate online. I enthusiastically agreed.

Other prominent Asian-American business leaders joined the effort, including Yahoo cofounder Jerry Yang; Alibaba cofounder and executive vice chairman Joe Tsai; Peng Zhao, CEO of Citadel Securities; Joe Bae, copresident and co-COO of KKR; and Sheila Marcelo, cofounder of Care.com. Sonal Shah, a friend who had worked in government, business, and academia, later came on board as the organization's first president.

In May 2021, this group launched the Asian American Foundation (TAAF), catalyzing more than $1.1 billion in new philanthropic resources for the AAPI community. That sum included $250 million drawn from a combination of their own contributions and pledges from companies like Walmart, Bank of America, Coca-Cola, Merck, and Verizon. To date, this is the biggest philanthropic effort ever mounted in the United States for an Asian-American cause.

TAAF has a simple but powerful mission: "To serve the Asian American and Pacific Islander community in their pursuit of belonging and prosperity that is free from discrimination, slander, and violence." I am proud that ADL worked quietly for a year before this organization's launch to support it with resources, staff, and training.

TAAF's founding exemplifies the kind of allyship and solidarity that ADL seeks to show and cultivate in others. All of us must step up to fight hate *everywhere,* not just when it's directed at us or our groups. But TAAF's existence also speaks to a new drive on the part of business leaders and companies to stand up against hate. These prominent members of the AAPI community understood the wisdom and necessity of investing time and money in their community to counter those who would cause harm to it. They're not waiting for government or the courts to solve the problem. There's a leadership vacuum, and the executives on the TAAF board are filling it.

If you run a business, I ask you to consider what *you* might do. What organizations might you join or found? How might you transform your business operations to further inclusivity and tolerance? You have the power to help keep society from rising farther up the pyramid. Your stakeholders expect it. Common decency demands it. I hope you'll join other courageous business leaders and rise to the challenge, helping to ensure a safe, peaceful, and prosperous society for future generations.

# EPILOGUE

The book of Exodus contains a memorable story about the prophet Moses, one that I've sometimes invoked in my speeches. An unremarkable shepherd, Moses is walking through the desert while tending a flock owned by his father-in-law when he comes upon an oddity — a bush that appears to be burning but whose flames aren't consuming any vegetation. Stopping to marvel at this "great sight," he hears a divine voice calling out to him from the bush: "Moses, Moses."

Moses doesn't run away in fright. He doesn't respond with a quizzical look and a mundane "Yes?" Rather, he says, *"He-ne-ni,"* which translates as "Here I am." It's a bold way to assert his presence before God. It's as if Moses is stepping forward and showing up, ready for whatever will come of the encounter.

God has big plans for Moses. He wants this shepherd to rectify an epic injustice and save an entire people from enslavement. "I have surely seen the affliction of My people that are in Egypt," God says, "and have heard their cry by reason of their taskmasters; for I know their pains."

Moses harbors doubts. He is a simple man and one racked by an insecurity that would define him throughout his life. How could he possibly succeed in freeing God's people? But after some reassurances from the Lord, he accepts the call to duty. It all starts with him showing up and loudly and unequivocally saying, *Here I am.*

When it comes to today's plague of bigotry and incivility, we should all draw inspiration from Moses's response. In this book, I've sought to raise the alarm about the problem of hate and its spread in the United States and around the world. Biases and slurs might seem inconsequential, but if left unchecked, they can progress to sustained harassment, systemic discrimination, acts of violence, mass unrest, and, ultimately, the kind of brutal conflagration that we all shudder to imagine. As tempting as it might be to ignore hate or treat it as someone else's problem, we do that at our peril. The prospect of widespread violence and genocide seems improbable, even unthinkable — until it isn't.

America has been moving perilously up the Pyramid of Hate in recent years. Hate crimes have soared, including horrific acts of mass violence aimed at Jews, AAPIs, Blacks, Latinos, LGBTQ people, Muslims, and others. Some in the Jewish community speak of the United States as becoming Europeanized, meaning that pervasive intolerance is making it increasingly difficult for American Jews — like European Jews before them — to express their identities openly. Based on what I've seen over the past few years, with schools being fortified, synagogues bolstering their security, and Jews hiding outward signs of their identity in public places, I can't help but agree.

And the issues are not unique to the United States. It feels like global society is coming apart, riven by rampant polarization, sectarian unrest, and increasing dehumanization. I fear that we're at a make-or-break moment. We can act now to tamp down bigotry and intolerance before they escalate into chaos. Or we can sit back and let it subsume us. The choice is ours.

Years ago, when the ADL first asked me to interview for the job of CEO, I wasn't sure if I was up to the task. Based on my résumé, I felt that I was not the most qualified person to lead this vaunted civil rights organization, literally an American institution. I had other career plans in mind that would have returned me to the business world and put me and my family on the opposite side of the country. But after consider-able reflection and consultation with several close friends, I felt ready to answer the call—to say, in essence, *Here I am*. Many factors led to that decision, but foremost among them were my memories of my grandfa-ther's trauma and the extermination of European Jewry, the knowledge of my wife's trauma and the exile of Mizrahi Jewry, and my determina-tion that such catastrophes should never happen again.

Now it's your turn to say, *Here I am*.

If you have relatives or friends who have fallen victim to hate, do it for them. But do it also for yourself and your children. Think of the world *you* want to leave to future generations. We're privileged today to live in democracies that pursue, albeit imperfectly, cherished values like equality, the rule of law, and self-determination. But our thriving so-cieties didn't just materialize on their own. The freedoms we enjoy are hard won, delivered through the sacrifice of countless others in prior generations who stepped up and said, *Here I am*.

You might think people are powerless against the forces of hate, but rest assured, we're strong. When we all decide to join the fight and bend the arc of the moral universe toward justice, we can accomplish ex-traordinary things. In this book I've tried to describe steps all of us can take in our roles as citizens, parents, educators, activists, business exec-utives, political leaders, and people of faith. Again and again, we at the ADL have seen entire communities come together to reaffirm common standards of decency and inclusivity in the face of bigotry. We've seen individuals affected by hate who, instead of cowering in fear, decided to take a stand and do something. We've even seen former extremists courageously overcome their hate, come to grips with their misdeeds, and work to help others.

There are countless inspiring stories of courage I might have told in these pages were it not for a lack of space. I might have described the Haitian refugee who, after being bullied, helped change the climate at her school. And the university baseball coach who, after hearing a player use a slur, took steps to educate his team about prejudice. And the police captain from a small city who arranged for his entire department to receive training in implicit bias. And the restaurant owner who, upon encountering an organized group of Holocaust deniers in his establishment, ejected all of them and subsequently donated the exact amount of their bill to the ADL.

What's your story of courage? No matter your age, social status, or political beliefs, you can make a profound difference in your family, community, and beyond. Examine your own biases. Educate yourself and others. Serve as an ally to others experiencing harassment. Affirm the values of tolerance and inclusivity. Advocate for social justice. Perform small acts of caring and compassion.

Yes, the unthinkable *could* happen here. But it doesn't have to. And if we all muster the strength right now to stand up and say, *Here I am,* it won't.

# ACKNOWLEDGMENTS

Writing a book is a team sport, and as a result I owe many debts of gratitude.

First, to the team that made this book possible: My literary agent Howard Yoon from the Ross Yoon Agency, whose creativity is matched only by his patience, pushed me to write this book from the very beginning. Without his advocacy, there would have been no book. Dan Gerstein made that *shidduch* and I'm deeply grateful for that. Seth Schulman and everyone at Providence Word & Thought contributed enormously to the writing process; Seth personally was my partner every step of the way. Alex Littlefield at Mariner Books believed in the idea for this book and pushed me to use the Pyramid of Hate as a framework. I appreciated the close read and smart suggestions from him and his team throughout the process.

My gratitude goes out as well to all those who sat for interviews: Imam Abdullah Antepli, Andy Berke, Greg Ehrie, Allison Goodman-Padilla, Averi Kaplowich, Adna Karamehic-Oates, Steven Levitsky, Rabbi Hazzan Jeffrey Myers, Damien Patton, Oren Segal, Jinnie Spiegler, Gregory Stanton, Robert Trestan, Barbara Walter, and two interviewees whom I will keep anonymous. I also wish to thank those friends who generously read early versions of my manuscript and offered helpful and sometimes hard feedback: Yossi Klein Halevi, Brian

Hooks, Jack Leslie, Batya Ungar-Sargon, Ken Baer, and Eboo Patel. Ken Jacobson, Jessica Stallone, Justin Finkelstein, and Ryan Greer from my team at ADL also gave detailed reads and provided key research that made this book better.

More broadly I wish to thank everyone at ADL. If I had the space, I would recognize every single one of our 362 full-time employees, each of our 450 part-time trainers and contractors, all 335 members of our National Commission; the 120 members of our Global Leadership Council; the 1,500 members of the Glass Leadership Institute community; the 45 trustees of the ADL Foundation; the 18 individuals who serve with me on the ADL board of directors; the tens of thousands of people who support us each year with their generous contributions; and the millions of students, teachers, HR managers, elected officials, law enforcement officers, and so many others who rely on our content on a daily basis. Unfortunately, I can't name all of these people, but there are a few among them who deserve special recognition.

First, thank you to my executive team at the ADL Community Support Center (CSC), including Adam Neufeld, Anat Kendal, Eileen Hershenov, Emily Bromberg, Fred Bloch, George Selim, Sharon Nazarian, Steve Sheinberg, and Tom Ruderman. I also want to recognize the members of our full CSC senior team, including Larry Chertoff, Greg Ehrie, Steve Freeman, Deb Lehrer, Deb Leipzig, Dan Roberti, Oren Segal, Max Sevillia, and Dave Sifry. Nike Irvin has earned my deepest appreciation for her leadership and grace in building and guiding the Civil Society Fellowship. Cliff Schechter made my transition into this job possible back in 2015 and has been a trusted colleague and honest broker ever since. Several members of the CSC team who have left ADL over the years were essential to my ADL journey: Betsaida Alcantara, Amy Blumkin, Evan Bernstein, Shari Gersten, Todd Gutnick, Lonnie Nasatir, and Marvin Rappaport.

I want to give a special acknowledgment to the aforementioned

Ken Jacobson. He is the heart and soul of ADL, a fifty-year veteran of this organization. As others operated out in front and earned accolades over the years, Ken kept his head down, stayed in the background, and just did the work. He has made me a better leader, keeping me honest and serving as the conscience of this agency. Whether or not you know his name, Ken has been a huge part of Jewish communal life for decades. All of us have benefited from his quiet brilliance and countless sacrifices.

The magic of ADL isn't what happens at our headquarters but rather what happens in the field, where incidents are handled, children are educated, legislation is passed, arrests are made, and the world is changed. To that end, special recognition goes out to our regional leadership, starting with our divisional vice presidents: Cheryl Drazin, Doron Ezickson, Allison Goodman-Padilla, and Robert Trestan. I also want to acknowledge the dedicated professionals managing our regional offices in the U.S. and overseas: Jeff Abrams, Jolie Brislin, Seth Brysk, Miri Cypers, Sarah Emmons, Linsday Bach Friedmann, Tammy Gillies, Steve Ginsberg, David Goldenberg, Shira Goodman, Rene Lafair, Peter Levi, Scott Levin, Dan Meisel, Gary Nachman, Carolyn Normandin, Carole Nuriel, James Pasch, Scott Richman, and Mark Toubin.

A few other ADL staffers no longer work at the agency but were huge influences on me. Sally Greenberg and Cheryl Cutler Azaire supervised me as a twenty-year-old college intern at ADL's office in downtown Boston. Lenny Zakim (z"l), the New England office's legendary director, allowed me to be there in the first place, enabling me to take the first step on my professional journey. He left this world far too young, but his memory lives on through the countless people influenced by his prophetic leadership.

I also humbly thank the ADL leaders who came before me. While I didn't work directly with any of them, I still stand on their shoulders. These individuals include our founder, Sigmund Livingston (z"l), whose audacious vision still guides ADL; Leon Lewis (z"l), whose bravery prompted him to find and fight Nazis in America; Richard Gutstadt (z"l), whose professionalism presaged the modern ADL; Ben Epstein (z"l), whose extraordinary courage enabled ADL to live up to its mission of fighting for justice and fair treatment to all; Nathan Perlmutter (z"l), whose charisma and cleverness put this agency on strong financial footing; and Abe Foxman, whose moral authority demanded attention and dominated the landscape for so long.

The generosity of our donors has enabled my success at ADL. Over the years, hundreds of thousands of people have made charitable contributions to this organization, and I feel a personal responsibility to each and every one of them. Several donors have made transformative gifts since I came to ADL. These individuals took a chance on me and have stuck by me even if they haven't agreed with every decision. To Jeanie and Jonathan Lavine, Marc Rowan, Daniel Loeb, Kathryn and James Murdoch, Lester Crown, and David Tepper: I'm humbled by your philanthropy and thankful for your loyalty.

Finally, ADL matters because of our volunteer leaders, and I've benefited enormously from their counsel and guidance. When I was an ADL intern, David Bunis and Rikki Kleiman gave me life-shaping advice that they probably don't even remember. You can credit (or blame!) Barry Curtiss-Lusher for recruiting me in 2013, when he served as national chair. Marvin Nathan succeeded him and was an invaluable adviser and confidant as we did many hard things together during his tenure. Esta Epstein has been a dear friend and a pillar of strength as the inaugural chair of our revitalized board of directors. And I have learned much from the other former national chairs I have known: Ken Bialkin (z"l), Burt Levinson, Mel Salberg (z"l), David Strassler, Howard Berkowitz, Glen Tobias, Barbara Balser, Glen Lewy, and Bob Sugarman. I have no doubt that our incoming board chair, Ben Sax, will take his place among this pantheon of giants.

Of course, my life began long before ADL. My grandparents played a formative role during my childhood, particularly Bernard Greenblatt and the daring he showed in the face of Nazi Germany and in the most horrifying circumstances. I'm also inspired by my mother- and father-in-law, Lili and Mansour Keypour, who summoned similar strength to rescue their daughters and themselves from the brutal grip of the Islamic Republic of Iran.

A series of mentors believed in me over the years. I spent two tours of duty in Washington, DC, and had the good fortune to work for President Clinton and President Obama, both of whom generously gave me the opportunity to serve my country. Other incredible mentors included Michael Whouley, Rob Stein, Jonathan Silver, Ira Magaziner, David Rothkopf, Cecilia Muñoz, and John Gomperts. While serving on President Obama's team, I was lucky to work with some extremely skilled colleagues who were a huge part of what I accomplished. Thanks to Carlos Monje, Tom Kalil, Danielle Gray, Racquel Russell, Michael Smith, and Broderick Johnson for their collaboration and friendship. I also still cherish my SICP team: Noemie Levy, Jess Yuen, Dave Wilkinson, Annie Donovan, and Rafael Lopez.

In the business world, many people contributed to my success, but I want to acknowledge just a few. David Rosenblatt and Russ Mann guided me at Realtor.com and became dear friends. Peter Thum had the original vision for Ethos Water, and together we founded Ethos Brands, but we truly were buoyed by so many people, starting with Ari Engleberg and Steven Koltai, who made crucial angel investments; Seth Goldman, Peter Meehan, and Gary White, who provided invaluable advice along the way; Michael Segal, Chris Anderson, and Michael Besancon, each of whom gave us big breaks and key exposure; Pam and Pierre Omidyar, along with Mike Mohr, who believed in us and provided the growth capital that allowed us to scale; and Howard Schultz, who recognized our potential and had the vision to acquire Ethos and distribute the brand through Starbucks stores across North America. I also have the deepest appreciation for Jim Donald, Gerry Lopez, and Colleen Chapman along with the hundreds of thousands of Starbucks partners over the years who have imbued Ethos Water with their passion, not only making it one of the most visible bottled-water brands in America but also helping so many children around the world get clean water.

Since my time at Starbucks, I've been lucky to work with more amazing people: Michael Madnick and the team at the UN Foundation; Ben Goldhirsh and all the remarkably talented folks at GOOD; Katie Jacobs Stanton, Chris DiBona, Adam Sah, and all the Googlers who helped make All for Good the largest database of volunteer opportunities on the web; as well as Susan Nesbitt, David Eisner, Michelle Nunn, Arianna Huffington, Craig Newmark, Bobbi Silten, Sonny Jandial, and the other investors, volunteers, and supporters who assisted AFG and enabled its success through its acquisition by Points of Light.

Few experiences teach you as much as actually being a teacher. I appreciate Judy Olian, now president of Quinnipiac University, who recruited me to join the faculty at UCLA Anderson when she was its dean and develop its first course on social entrepreneurship long before the topic was *en vogue*. Geoff Garrett brought me to the Wharton School when he served as its dean and gave me a similar opportunity. Vice Dean Katherine Klein helped me make an impact on campus.

I've been lucky to serve on several nonprofit boards led by innovative, visionary founders. I absorbed many lessons from their entrepreneurial, impassioned leadership: Gary White at Water.org; Darell Hammond at Kaboom!; Fred Swaniker at the African Leadership Academy; and Mike Mallory at the Ron Brown Scholar Program. And I have learned so much from my fellow board members at the Asian American Foundation: Joe Bae, Li Lu, Sheila Marcelo, Joe Tsai, Jerry Yang, and Peng Zhao.

Although many people who have been part of my life are no longer with us, no loss affected me as much as the plane crash on April 3, 1996, that took the lives of thirty-four people, including Commerce Secretary Ron Brown, Assistant Secretary Chuck Meissner, and other dear colleagues and close friends: Duane Christian, Adam Darling, Gail Dobert, Kathy Kellogg, Kathryn Hoffman, Bill Morton, Lawrence Payne, and Naomi Warbasse. The world has been poorer since that terrible day. May their memories always be a blessing.

If I had a personal board of directors, many of the people I've already mentioned would fill its seats, as would Anthony Romero, Sacha Baron Cohen, Maninder Dhillon, Cheryl Dorsey, Derrick Johnson, Daniel Lubetzky, Bill Moses, Seth Oster, Sonal Shah, Bret Stephens, and Rabbi David Wolpe. I'm better every day for their honesty and friendship.

Finally, I want to thank my family: My parents, who shaped me with their guidance and values; my brother, whose sense of humor keeps me smiling; and my three boys, whom I love more than anything. And my wife, Marjan, is not only my rock but my closest adviser, best friend, and the love of my life. I could not do anything were it not for her tremendous, unwavering support.

# NOTES

## INTRODUCTION

*page*

3    *a brutal episode of anti-Jewish hate:* Matt Lebovic, "The ADL and KKK, Born of the Same Murder, 100 Years Ago," *Times of Israel,* May 27, 2013, https://www.timesofisrael.com/the-adl-and-kkk-born-of-the-same-murder-100-years-ago/.

*"secure justice and fair treatment":* Anti-Defamation League Founding Charter, Anti-Defamation League, October 1913, https://www.adl.org/who-we-are/excerpt-of-the-anti-defamation-league-founding-charter.

4    *a doubling of antisemitic incidents:* "Anti-Semitic Incidents Remained at Near-Historic Levels in 2018; Assaults Against Jews More Than Doubled," Anti-Defamation League, April 30, 2019, https://www.adl.org/news/press-releases/anti-semitic-incidents-remained-at-near-historic-levels-in-2018-assaults.

*in the past* four decades: "Antisemitic Incidents Hit All-Time High in 2019," Anti-Defamation League, May 12, 2020, https://www.adl.org/news/press-releases/antisemitic-incidents-hit-all-time-high-in-2019.

*antisemitic imagery was on startling display:* Elana Schor, "Anti-Semitism Seen in Capitol Insurrection Raises Alarms," *U.S. News and World Report,* January 13, 2021, https://www.usnews.com/news/politics/articles/2021-01-13/anti-semitism-seen-in-capitol-insurrection-raises-alarms.

*"cops and Zionists":* See @MPower_Change, https://twitter.com/MPower_Change/status/1273347515765702656, June 17, 2020.

*regard the State of Israel:* Frank Newport, "American Jews, Politics, and Israel," Gallup, August 27, 2019, https://news.gallup.com/opinion/polling-matters/265898/american-jews-politics-israel.aspx.

5    *deployed rhetorical violence:* "Antisemitic Incidents at Anti-Israel Events and Actions Around the World," Anti-Defamation League, last updated June 1, 2021, https://www.adl.org/resources/fact-sheets/antisemitic-incidents-at-anti-israel-events-and-actions-around-the-world.

*real-world violence:* "Following Start of Mideast Violence, Antisemitic Incidents More Than Double in May 2021 vs. May 2020," Anti-Defamation League, June 7, 2021; see also *Center on Extremism: As It Happens* (blog), Anti-Defamation League, May 12, 2021, https://www.adl.org/extremismblog.

*hatred starts with the Jews:* Deborah Lipstadt, *Antisemitism Here and Now* (New York: Schocken, 2019), xi.

*7,314 hate crimes:* Michael Balsamo, "Hate Crimes in US Reach Highest Level in More Than a Decade," AP News, November 16, 2020, https://apnews.com/article/hate-crimes-rise-fbi-data-ebbcadca8458aba96575da905650120d.

*hate crimes against Asian-Americans:* Masood Farivar, "Hate Crimes Targeting Asian Americans Spiked by 150% in Major US Cities," *VOA News,* March 2, 2021, https://www.voanews.com/usa/race-america/hate-crimes-targeting-asian-americans-spiked-150-major-us-cities.

*44 percent of Americans:* "Online Hate and Harassment Report: The American Experience 2020," Anti-Defamation League, June 2020, https://www.adl.org/online-hate-2020.

6   *the foreword for a book:* "Corbyn Found to Have Written Foreword for Book Claiming Jews Control Banks," *Times of Israel,* May 1, 2019, https://www.timesofisrael.com/corbyn-found-to-have-written-foreword-for-book-claiming-jews-control-banks/.

*"having lived in this country":* Daniel Sugarman, "Jeremy Corbyn: 'Zionists' Have 'No Sense of English Irony Despite Having Lived Here All Their Lives,'" *Jewish Chronicle,* August 23, 2018, https://www.thejc.com/news/uk/jeremy-corbyn-zionists-have-no-sense-of-english-irony-1.468795; "Jeremy Corbyn's 2013 Remarks on Some Zionists Not Understanding English Irony," *Guardian,* August 24, 2018, https://www.theguardian.com/politics/video/2018/aug/24/jeremy-corbyns-2013-remarks-on-some-zionists-not-understanding-english-irony-video.

*critical mass of British Jews:* Lee Harpin, "Nearly 40 Per Cent of British Jews Would 'Seriously Consider' Emigrating if Corbyn Became PM," *Jewish Chronicle,* September 5, 2018, https://www.thejc.com/news/uk/nearly-40-per-cent-of-british-jews-would-seriously-consider-emigrating-if-corbyn-became-pm-1.469270.

*latent psychological impulse:* The underlying psychology of hate is complex, and some resist seeing hate as a matter of intrinsic drives. For additional insight, please see Aaron T. Beck, *Prisoners of Hate: The Cognitive Basis of Anger, Hostility, and Violence* (New York: HarperCollins, 1999), and Robert J. Sternberg and Karin Sternberg, *The Nature of Hate* (Cambridge: Cambridge University Press, 2008), especially chapter 2.

7   *increasing diversity of the American population:* William Frey, "The Nation Is Diversifying Even Faster Than Predicted, According to New Census Data," Brookings, July 1, 2020, https://www.brookings.edu/research/new-census-data-shows-the-nation-is-diversifying-even-faster-than-predicted/.

*had a sordid history:* German Lopez, "Donald Trump's Long History of Racism, from the 1970s to 2020," *Vox,* updated August 13, 2020, https://www.vox.com/2016/7/25/12270880/donald-trump-racist-racism-history.

8   *asked a Jewish candidate:* "Stanford Student Senate Candidate Files Complaint After Being Questioned About Jewish Faith," *Haaretz,* April 14, 2015, https://www.haaretz.com/jewish/stanford-student-senate-candidate-asked-about-jewish-faith-1.5350811.

9   *identifiably Jewish individual:* Cade Johnson, "UC Berkeley Lecturer Apologizes for Anti-Semitic Retweet," November 27, 2017, https://www.dailycal.org/2017/11/27/uc-berkeley-lecturer-apologizes-anti-semitic-retweet/.

*vehemently antisemitic messages:* Gabriel Greschler, "UC Merced Professor Deletes Antisemitism-Laden Twitter Account," *Jewish News of Northern California,* December 21, 2020, https://www.jweekly.com/2020/12/21/uc-merced-professor-deletes-antisemitism-laden-twitter-account/.

*75 percent rise in anti-Jewish incidents:* Lexi Lonas, "ADL Reports '75 Percent Surge' in Antisemitic Attacks in Last Two Weeks," *Hill,* May 27, 2021, https://thehill.com/blogs/blog-briefing-room/news/555843-adl-reports-75-percent-surge-in-antisemitic-attacks-in-last-two.

*majority of Jews:* "Survey of American Jews Since Recent Violence in Israel," Anti-Defamation League, June 9, 2021, https://www.adl.org/blog/survey-of-american-jews-since-recent-violence-in-israel.

*attacked a group of Jewish men:* "Police Investigate Possible Jewish Hate Crime Attack at Beverly Grove Restaurant," CBS Los Angeles, May 19, 2021, https://losangeles.cbslocal.com/2021/05/19/police-investigate-possible-jewish-hate-crime-attack-at-beverly-grove-restaurant/.

*shouting antisemitic invective:* Sara Marcus, "Queens Man Attacked in Apparent Hate Crime Outside Shul," *Hamodia,* April 20, 2021, https://hamodia.com/2021/04/20/queens-man-attacked-apparent-hate-crime-outside-shul/.

*threatened to rape:* Carli Teproff, "'Die Jew.' Jewish Family Visiting South Florida Harassed While Walking in Bal Harbour," *Miami Herald,* May 21, 2021, https://www.miamiherald.com/news/local/community/miami-dade/aventura/article251584103.html.

*unable to offer clear, cogent condemnations:* Missouri congresswoman Cori Bush, for instance, tweeted: "The work of dismantling antisemitism, anti-Blackness, Islamophobia, anti-Palestinian racism, and every other form of hate is OUR work," @CoriBush, May 22, 2021. Rather than simply standing with victimized Jewish students, the president of Rutgers University apologized for initially calling out the surge of antisemitism, expressing his regret for alienating "Palestinian community members"; see Brittany Bernstein, "Rutgers Chancellor Apologizes for Condemning Anti-Semitic Attacks," *National Review,* May 28, 2021, https://www.nationalreview.com/news/rutgers-university-new-brunswick-chancellor-issues-apology-for-message-condemning-anti-semitic-attacks/. See also Melissa Block, "Antisemitism Spikes, and Many Jews Wonder: Where Are Our Allies?," NPR.org, June 7, 2021, https://www.npr.org/2021/06/07/1003411933/antisemitism-spikes-and-many-jews-wonder-where-are-our-allies.

10   *"common respect for uniquely Anglo-Saxon political traditions":* Daniella Diaz, "Marjorie Taylor Greene Scraps Planned Launch of Controversial 'America First' Caucus Amid Blowback from GOP," CNN, April 18, 2021, https://www.cnn.com/2021/04/17/politics/marjorie-taylor-greene-america-first-caucus/index.html.

13   *40 percent of the U.S. population:* Frederic Cople Jaher, *The Jews and the Nation: Revolution, Emancipation, State Formation, and the Liberal Paradigm in America and France* (Princeton, NJ: Princeton University Press, 2002), 230.

*11 percent of the population:* "Antisemitic Attitudes in the U.S.: A Guide to ADL's Latest Poll," Anti-Defamation League, January 2020, https://www.adl.org/survey-of-american-attitudes-toward-jews. For methodological reasons, the comparison between 1964 and the 1930s is inexact but, in my view, still generally valid.

*"the social responsibility of business":* Milton Friedman, "A Friedman Doctrine—the Social Responsibility of Business Is to Increase Its Profits," *New York Times,* September 13, 1970,

https://www.nytimes.com/1970/09/13/archives/a-friedman-doctrine-the-social-responsibility-of-business-is-to.html.

14   *the world's largest companies:* Brian Groom, "A Third of Start-Ups Aim for Social Good," *Financial Times,* June 14, 2018, https://www.ft.com/content/d8b6d9fa-4eb8-11e8-ac41-759eee1efb74.

*"The arc of the moral universe":* For the original quote, see, for instance, Martin Luther King Jr., "Sermon at Temple Israel of Hollywood," *American Rhetoric,* https://www.americanrhetoric.com/speeches/mlktempleisraelhollywood.htm.

15   *quoting Voltaire:* Voltaire, we should note, himself invested in slave ships and was a vicious antisemite; he wrote of the Jews: "They are, all of them, born with raging fanaticism in their hearts, just as the Bretons and the Germans are born with blond hair." See Nabila Ramdani, "Voltaire Spread Darkness, Not Enlightenment. France Should Stop Worshipping Him," *Foreign Policy,* August 31, 2020, https://foreignpolicy.com/2020/08/31/voltaire-spread-darkness-not-enlightenment-france-should-stop-worship-paris-statue-racism-black-lives-matter/.

## 1. HATE GONE MAINSTREAM

19   *acts of mass violence:* Agence France-Presse, "Anti-Semitic Attacks in Europe in Recent Years," *Times of Israel,* January 7, 2019, https://www.timesofisrael.com/anti-semitic-attacks-in-europe-in-recent-years/; Kim Willsher, "Paris Court Hears How Kosher Supermarket Attacker Killed Four," *Guardian,* September 21, 2020, https://www.theguardian.com/world/2020/sep/21/paris-court-kosher-supermarket-attacker-amedy-coulibaly-killed-four-charlie-hebdo.

21   *"I can't sit by":* See "Deadly Shooting at Pittsburgh Synagogue," Anti-Defamation League, October 27, 2018, https://www.adl.org/blog/deadly-shooting-at-pittsburgh-synagogue.

22   *white Stars of David:* Associated Press, "Pittsburgh Shooting Suspect Indicted on 44 Counts, Including Hate Crimes," PBS.org, October 31, 2018, https://www.pbs.org/newshour/nation/pittsburgh-shooting-suspect-indicted-on-44-counts-including-hate-crimes.

24   *I asked Schubert:* I recounted my experiences that day in Pittsburgh in a written statement; see Jonathan Greenblatt, "Standing Together to Support the Community," Anti-Defamation League, November 5, 2018, https://www.adl.org/blog/standing-together-to-support-the-community.

25   *killing a woman:* "Poway Attack Illustrates Danger Right-Wing Extremists Pose to Jews, Muslims," Anti-Defamation League, May 2, 2019, https://www.adl.org/blog/poway-attack-illustrates-danger-right-wing-extremists-pose-to-jews-muslims; Stephanie K. Baer, "The Poway Synagogue Shooter Planned a Larger Attack but Stopped Firing After His Rifle Jammed," *BuzzFeed,* April 30, 2019, https://www.buzzfeednews.com/article/skbaer/poway-synagogue-shooter-rifle.

*injuring twenty-six:* "The El Paso Attack, One Year Later: Extremist Threat Remains High," Anti-Defamation League, July 31, 2020, https://www.adl.org/blog/the-el-paso-attack-one-year-later-extremist-threat-remains-high.

*Black Hebrew Israelite community:* As the ADL has noted, "The Black Hebrew Israelite (BHI) movement is a fringe religious movement that rejects widely accepted definitions of Judaism and asserts that people of color are the true children of Israel. Followers . . .

believe that Blacks, Hispanics, and Native Americans are the descendants of the Twelve Tribes of Israel." See "Extremist Sects within the Black Hebrew Israelite Movement," Anti-Defamation League, https://www.adl.org/resources/backgrounders/extremist-sects-within-the-black-hebrew-israelite-movement.

*harbored antisemitic views:* "ADL Report: Right-Wing Extremists Killed 38 People in 2019, Far Surpassing All Other Murderous Extremists," Anti-Defamation League, February 26, 2020, https://www.adl.org/news/press-releases/adl-report-right-wing-extremists-killed-38-people-in-2019-far-surpassing-all; David Porter and Michael Sisak, "Official: Jersey City Attack Was 'Fueled' by Anti-Semitism," APNews.com, December 12, 2019, https://apnews.com/article/20dbf0854c6fb812fcf4b979986492d7.

*searched online for Zionist temples:* Michael Gold and Benjamin Weiser, "Suspect in Monsey Stabbings Searched Online for 'Hitler,' Charges Say," *New York Times,* December 30, 2019, https://www.nytimes.com/2019/12/30/nyregion/jewish-attacks.html.

25    *killed fifty-one worshippers:* Eleanor Ainge Roy and Charlotte Graham-McLay, "Christchurch Gunman Pleads Guilty to New Zealand Mosque Attacks That Killed 51," *Guardian,* March 25, 2020, https://www.theguardian.com/world/2020/mar/26/christchurch-shooting-brenton-tarrant-pleads-guilty-to-new-zealand-mosque-attacks-that-killed-51.

26    *Deadly attacks by white supremacists:* Weiyi Cai and Simone Landon, "Attacks by White Extremists Are Growing. So Are Their Connections," *New York Times,* April 3, 2019, https://www.nytimes.com/interactive/2019/04/03/world/white-extremist-terrorism-christchurch.html.

*The Pulse nightclub in Orlando:* The perpetrator, Omar Mateen, claimed fidelity to ISIS, although his precise motivations remain unclear. See, for instance, Paul Brinkmann, "Pulse Gunman's Motive: Plenty of Theories, but Few Answers," *Orlando Sentinel,* June 4, 2017, https://www.orlandosentinel.com/news/pulse-orlando-nightclub-shooting/os-pulse-omar-mateen-motive-20170512-story.html.

*so-called Zoom-bombing:* Sebastian Meineck and Paul Schwenn, "'Zoom Bombers' Are Still Blasting Private Meetings with Disturbing and Graphic Content," *Vice,* June 10, 2020, https://www.vice.com/en/article/m7je5y/zoom-bombers-private-calls-disturbing-content.

*white supremacist and other Far Right groups:* Robert O'Harrow Jr., Andrew Ba Tran, and Derek Hawkins, "The Rise of Domestic Extremism in America," *Washington Post,* April 12, 2021, https://www.washingtonpost.com/investigations/interactive/2021/domestic-terrorism-data/.

*"That increase is nothing":* Gregory Ehrie, interview with the author, March 2, 2021.

27    *HEAT map:* "ADL H.E.A.T. Map," Anti-Defamation League, https://www.adl.org/education-and-resources/resource-knowledge-base/adl-heat-map. Please also consult our antisemitic incident tracker, available at https://www.adl.org/education-and-resources/resource-knowledge-base/adl-tracker-of-antisemitic-incidents?field_incident_location_state_target_id=All&page=2.

*white-nationalist ideology:* Eric K. Ward, "Skin in the Game: How Antisemitism Animates White Nationalism," Political Research Associates, June 29, 2017, https://www.politicalresearch.org/2017/06/29/skin-in-the-game-how-antisemitism-animates-white-nationalism.

28    *gave speeches at white-supremacist events:* Robert Costa, "Trump Speechwriter Fired Amid Scrutiny of Appearance with White Nationalists," *Washington Post,* August 19, 2018, https://www.washingtonpost.com/politics/trump-speechwriter-fired-amid-scrutiny-

of-appearance-with-white-nationalists/2018/08/19/f5051b52-a3eb-11e8-a656-943eefab5daf_story.html.

*white-power signs:* Josh Delk, "White House Intern Appeared to Use White Power Sign in Photo with Trump: Report," *Hill,* December 28, 2017, https://thehill.com/blogs/blog-briefing-room/366738-white-house-intern-used-white-power-sign-in-photo-with-trump.

*issued press credentials:* Michael M. Grynbaum, "Site That Ran Anti-Semitic Remarks Got Passes for Trump Trip," *New York Times,* January 26, 2020, https://www.nytimes.com/2020/01/26/business/media/trunews-white-house-press-credentials.html.

*compelled to resign:* Aaron Bandler, "USC Student VP Resigns, Says She Was Bullied for Being a Zionist," *Jewish Journal,* August 6, 2020, https://jewishjournal.com/featured/319981/usc-student-vp-resigns-says-she-was-bullied-for-being-a-zionist/.

*Sheldon Adelson:* Madeline Charbonneau, "Pink Floyd Co-Founder Roger Waters Accuses Sheldon Adelson of Being Trump's 'Puppet Master,'" *Daily Beast,* June 22, 2020, https://www.thedailybeast.com/pink-floyd-co-founder-roger-waters-accuses-sheldon-adelson-of-being-trumps-puppet-master.

*openly trafficked:* In one tweet, for instance, Peled wrote, "#Zionism is a sign of dual loyalty at best, at worst you are actually paid agents working for a foreign, racist government"; @mikopeled, Twitter, April 8, 2019, https://twitter.com/mikopeled/status/1115254243328499712?lang=en.

29    *classic antisemitic tropes:* Izabella Tabarovsky, "We Soviet Jews Lived Through State-Sponsored Anti-Zionism. We Know How It Is Weaponized," *Forward,* March 7, 2019, https://forward.com/opinion/420508/the-ussr-was-famous-for-state-sponsored-anti-zionism-is-america-heading-in/.

*From Iraq to Egypt:* Sarah Ehrlich, "Farhud Memories: Baghdad's 1941 Slaughter of the Jews," BBC, June 1, 2011, https://www.bbc.com/news/world-middle-east-13610702; George E. Gruen, "The Other Refugees: Jews of the Arab World," Jerusalem Letter for Public Affairs, June 1, 1988, https://www.jcpa.org/jl/jl102.htm.

*Tulsa race massacre of 1921:* See, for instance, "Tulsa Race Massacre," History.com, last updated June 1, 2021, https://www.history.com/topics/roaring-twenties/tulsa-race-massacre.

*Chinese massacre of 1871:* Shelby Grad, "The Racist Massacre That Killed 10% of L.A.'s Chinese Population and Brought Shame to the City," *Los Angeles Times,* March 18, 2021, https://www.latimes.com/california/story/2021-03-18/reflecting-los-angeles-chinatown-massacre-after-atlanta-shootings.

30    *Islamic State's earlier technique:* Didier Lauras, "Shared Tactics Fan Flames for Jihadists and White Supremacists," *Barron's,* February 5, 2021, https://www.barrons.com/news/shared-tactics-fan-flames-for-jihadists-and-white-supremacists-01612537505.

*in direct contact with:* Cai and Landon, "Attacks by White Extremists Are Growing."

*"White Jihad":* Lauras, "Shared Tactics."

*"It's easier than you think":* Philip Murphy, Sheila Oliver, and Jared Maples, "White Supremacist Extremists Exploit Jihadist Tactics," State of New Jersey, Office of Homeland Security and Preparedness, December 16, 2019, https://www.njhomelandsecurity.gov/analysis/white-supremacist-extremists-exploit-jihadist-tactics.

*"the culture of martyrdom":* Murphy, Oliver, and Maples, "White Supremacist Extremists."

*"Thinking that they shared":* Greg Walters and Tess Owen, "2 Boogaloos Bois Were Just

Busted for Trying to Sell Arms to Hamas," *Vice,* September 4, 2020, https://www.
vice.com/en/article/akz9jp/2-boogaloos-bois-were-just-busted-for-trying-to-sell-
arms-to-hamas.

31  *a terrorist group:* Amanda Coletta, "Canada Declares the Proud Boys a Terrorist Group,"
*Washington Post,* February 3, 2021, https://www.washingtonpost.com/world/the_
americas/canada-proud-boys-terrorist-capitol-siege/2021/02/03/546b1d5c-6628-
11eb-8468-21bc48f07fe5_story.html.
*"a social media phenomenon":* Oren Segal, interview with the author, March 2, 2021.
The ADL is hardly the only organization that links social media to the spread of hate;
see Zachary Laub, "Hate Speech on Social Media: Global Comparisons," Council on
Foreign Relations, June 7, 2019, https://www.cfr.org/backgrounder/hate-speech-
social-media-global-comparisons.

32  *nefarious global schemes:* See, for instance, "Leila Khaled Promotes Violent Resistance on
October ILPS Webinar," Anti-Defamation League, October 6, 2020, https://www.
adl.org/blog/leila-khaled-promotes-violent-resistance-on-october-ilps-webinar.
*"a lot of people":* Bess Levin, "Trump: 'A Lot of People Say' George Soros Is Funding
the Migrant Caravan," *Vanity Fair,* October 31, 2018, https://www.vanityfair.com/
news/2018/10/donald-trump-george-soros-caravan.

33  *"apocalyptic race war":* Daniel De Simone, "Neo-Nazi Group Led by 13-Year-Old Boy
to Be Banned," BBC, July 13, 2020, https://www.bbc.com/news/uk-53392036.

34  *"a forged ideology":* Dr. Osama Abuirshaid, remarks at the Islamic Center of South Florida, February 10, 2018, https://www.youtube.com/watch?v=9s0OoeQsyUw.
*their "double loyalty":* Dr. Osama Abuirshaid, remarks at the fifteenth annual convention of the Muslim American Society—Islamic Circle of North America
(MAS-ICNA), Chicago, Illinois, posted January 2017, https://www.youtube.com/
watch?v=-KqZg1xUp4I&t=870s.
*"a trojan horse to normalize Zionism":* Hatem Bazian, "Palestine Statement of Principles
for American Muslim Leaders and Communities," Facebook, May 20, 2021, https://
www.facebook.com/dr.bazian/posts/10112110255288133.
*Orthodox Jewish man raising his hands:* Caleb Parke, "UC Berkeley Jewish Students Say
University Protecting Lecturer Who Promotes Anti-Semitism," Fox News, December
6, 2017, https://www.foxnews.com/us/uc-berkeley-jewish-students-say-university-
protecting-lecturer-who-promotes-anti-semitism.

35  *personal account:* For a perspective on what happened, see Dexter Van Zile, "Seven Minutes of Hate Courtesy of SJP and UMASS Boston," *Times of Israel,* June 30, 2021,
https://blogs.timesofisrael.com/seven-minutes-of-hate-courtesy-of-sjp-and-umass-
boston/.
*"Your Party Platform remains the same":* "Re-Branding White Supremacy," Anti-Defamation League, December 16, 2016, https://www.adl.org/blog/re-branding-white-
supremacy.

36  *classic racist and antisemitic ideas:* "Alt Right: A Primer on the New White Supremacy,"
Anti-Defamation League, https://www.adl.org/resources/backgrounders/alt-right-a-
primer-on-the-new-white-supremacy.
*late Tony Martin:* "Schooled in Hate: Anti-Semitism on Campus," Anti-Defamation
League, https://www.adl.org/resources/reports/schooled-in-hate-anti-semitism-on-
campus.

38  *"another holiday to add to the calendar":* "'The Jews Should Count Themselves Lucky.'

Extremists React to Pittsburgh Synagogue Shooting," Anti-Defamation League, October 27, 2018, https://www.adl.org/blog/the-jews-should-count-themselves-lucky-extremists-react-to-pittsburgh-synagogue-shooting.

*Chabad synagogue in Poway, California:* "Multiple Threats to Synagogues in Wake of Tree of Life Shooting," Anti-Defamation League, December 24, 2018, https://www.adl.org/blog/multiple-threats-to-synagogues-in-wake-of-tree-of-life-shooting; "One Year After the Tree of Life Attack, American Jews Face Significant Threats," Anti-Defamation League, October 18, 2019, https://www.adl.org/blog/one-year-after-the-tree-of-life-attack-american-jews-face-significant-threats.

*vigil for the victims:* Bill O'Toole, "Allderdice Students Hold Vigil for the Tree of Life Victims," *NEXT Pittsburgh,* October 27, 2018, https://nextpittsburgh.com/latest-news/allderdice-students-hold-vigil-for-the-tree-of-life-victims/.

*moment of silence:* "Pittsburgh Steelers Hold Moment of Silence for Synagogue Shooting Victims," *Times of Israel,* October 28, 2018, https://www.timesofisrael.com/pittsburgh-steelers-hold-moment-of-silence-for-synagogue-shooting-victims/.

*businesses provided free goods and services:* Tracy Certo, "Music, Acupuncture, a Public Potluck: Local Businesses Offer Free Admission & Services to Comfort Grieving Pittsburgh," *NEXT Pittsburgh,* November 4, 2018, https://nextpittsburgh.com/latest-news/free-museum-admission-free-ice-cream-free-yoga-and-parking-pittsburghs-love-shines-through/; Melissa Rayworth, "Locally and Nationally, Support Keeps Pouring in for Pittsburgh and the Tree of Life Victims," *NEXT Pittsburgh,* October 31, 2018, https://nextpittsburgh.com/latest-news/locally-and-nationally-support-keeps-pouring-in-for-pittsburgh-and-the-tree-of-life-victims/.

*"will never be the same":* "Tree of Life — One Year Later," video, https://jwp.io/s/tYVwBHev.

*"There's a core decency":* "Tree of Life — One Year Later."

## 2. FROM MICROAGGRESSIONS TO GENOCIDE

40   *traded him away:* Donovan Dooley, "Racist Meyers Leonard Traded to OKC, Likely to Be Cut," *Deadspin,* March 18, 2021, https://deadspin.com/racist-meyers-leonard-traded-to-okc-likely-to-be-cut-1846502769; Sopan Deb and Kevin Draper, "Meyers Leonard Will Be Away from Heat 'Indefinitely' After Use of Anti-Semitic Slur," *New York Times,* March 9, 2021, https://www.nytimes.com/2021/03/09/sports/basketball/meyers-leonard-antisemetic-slur-.html.

*"seeking out people":* "Meyers Leonard, Already in Contact with Jewish Leaders, Getting the Ball Rolling," *TMZ Sports,* March 11, 2021, https://www.tmz.com/2021/03/11/meyers-leonard-already-in-contact-with-jewish-leaders-getting-the-ball-rolling/.

41   *three-quarters of American adults:* "The Trolls Are Organized and Everyone's a Target: The Effects of Online Hate and Harassment," Anti-Defamation League, October 2019, https://www.adl.org/trollsharassment#executive-summary-.

43   *"the Holocaust started with words":* "Survivor-Led Digital Campaign Launched by the Claims Conference Shines a Light on Hatred Before the Holocaust," Claims Conference, April 8, 2021, https://itstartedwithwords.org/about/. To highlight this connection between words and deeds, the Claims Conference launched an educational campaign called It Started with Words.

44   *the Pyramid of Hate:* For more on the Pyramid of Hate, see "Pyramid of Hate (en

Español)," Anti-Defamation League, https://www.adl.org/education/resources/
tools-and-strategies/pyramid-of-hate-en-espanol. The pyramid has been updated multiple times over the years, most recently in 2019. For an example of how we use it in
teaching, please see our document "The Escalation of Hate: Lesson for Middle and
High School Students," Anti-Defamation League, 2016, https://www.adl.org/sites/
default/files/documents/empowering-young-people-the-escalation-of-hate.pdf. I've
drawn on this document and others in writing this section.

45    *psychologists call "confirmation bias":* See Raymond Nickerson, "Confirmation Bias: A
      Ubiquitous Phenomenon in Many Guises," *Review of General Psychology* 2, no. 2 (June
      1998): 175–220.

      *implicit biases:* We can define *implicit biases* as "the unconscious attitudes, stereotypes and
      unintentional actions (positive or negative) toward members of a group merely be-
      cause of their membership in that group." See "Race, Perception and Implicit Bias,"
      Anti-Defamation League, https://www.adl.org/education/resources/tools-and-strat
      egies/table-talk/race-perception-and-implicit-bias; Jenée Desmond-Harris, "Implicit
      Bias Means We're All Probably at Least a Little Bit Racist," *Vox,* August 15, 2016,
      https://www.vox.com/2014/12/26/7443979/racism-implicit-racial-bias.

      *greater chance of being arrested:* "Criminal Justice Fact Sheet," NAACP, https://www.
      naacp.org/criminal-justice-fact-sheet.

      *lower-resourced schools:* "K-12 Disparity Facts and Statistics," UNCF, https://uncf.org/
      pages/k-12-disparity-facts-and-stats; "Black Students in the Condition of Education
      2020," National School Board Association, June 23, 2020, https://nsba.org/Perspectives/
      2020/black-students-condition-education.

      *dying of underlying conditions:* "Creating Equal Opportunities for Health," Centers for
      Disease Control and Prevention, last reviewed July 3, 2017, https://www.cdc.gov/
      vitalsigns/aahealth/index.html.

      *high levels of environmental pollution:* "Low-Income, Black Neighborhoods Still Hit Hard
      by Air Pollution," *ScienceDaily,* August 10, 2019, https://www.sciencedaily.com/
      releases/2019/08/190810094052.htm.

      *getting a bank loan:* Gene Marks, "Black-Owned Firms Are Twice as Likely to Be Re-
      jected for Loans. Is This Discrimination?," *Guardian,* January 16, 2020, https://www.
      theguardian.com/business/2020/jan/16/black-owned-firms-are-twice-as-likely-to-
      be-rejected-for-loans-is-this-discrimination.

      *disenfranchise Black voters:* Sanya Mansoor, "Georgia Has Enacted Sweeping Changes to
      Its Voting Law. Here's Why Voting Rights Advocates Are Worried," *Time,* March 26,
      2021, https://time.com/5950231/georgia-voting-rights-new-law/.

      *earn less than a man:* "Women Deserve Equal Pay," National Organization for Women
      Foundation, 2021, https://now.org/resource/women-deserve-equal-pay-factsheet/;
      "Barriers & Bias: The Status of Women in Leadership," American Association of Uni-
      versity Women, March 2016, https://www.aauw.org/resources/research/barrier-bias/.

46    *"the act or intent":* The creator of the modern term *genocide* defined it as "the destruction
      of a nation or an ethnic group," clarifying that it was "a coordinated plan of different
      actions aiming at the destruction of essential foundations of the life of national groups,
      with the aim of annihilating the groups themselves." See Adam Jones, *Genocide: A
      Comprehensive Introduction,* 2nd ed. (London: Routledge, 2011), 10. The United Na-
      tions formally designated genocide as a crime in 1948, conceiving it as "acts committed
      with intent to destroy, in whole or in part, a national, ethnical, racial, or religious
      group." Killing is one way that genocide can be carried out, according to the UN, but

it can also be carried out by inflicting serious harm, removing children from the group, preventing members of the group from reproducing, and inflicting conditions on the group for the purpose of destroying it; see Jones, *Genocide,* 13.

48    *"that had hosted"*: Ayal Feinberg, Regina Branton, and Valerie Martinez-Ebers, "Counties That Hosted a 2016 Trump Rally Saw a 226 Percent Increase in Hate Crimes," *Washington Post,* March 22, 2019, https://www.washingtonpost.com/politics/2019/03/22/trumps-rhetoric-does-inspire-more-hate-crimes/.

       *Bias against AAPIs has long existed:* Dale Minami, "The History of Anti-Asian Sentiment in the U.S.," NPR, March 18, 2021, https://www.npr.org/2021/03/18/978832077/the-history-of-anti-asian-sentiment-in-the-u-s.

       *3,800 hate crimes:* Kimmy Yam, "There Were 3,800 Anti-Asian Racist Incidents, Mostly Against Women, in Past Year," NBC News, March 16, 2021, https://www.nbcnews.com/news/asian-america/there-were-3-800-anti-asian-racist-incidents-mostly-against-n1261257.

       *85 percent spike:* "ADL Report: Anti-Asian Hostility Spikes on Twitter After President Trump's COVID Diagnosis," Anti-Defamation League, October 9, 2020, https://www.adl.org/news/press-releases/adl-report-anti-asian-hostility-spikes-on-twitter-after-president-trumps-covid.

49    *"There's a clear correlation":* Cady Lang, "Hate Crimes Against Asian Americans Are on the Rise. Many Say More Policing Isn't the Answer," *Time,* February 18, 2021, https://time.com/5938482/asian-american-attacks/. See also "Reports of Anti-Asian Assaults, Harassment and Hate Crimes Rise as Coronavirus Spreads," Anti-Defamation League, June 18, 2020, https://www.adl.org/blog/reports-of-anti-asian-assaults-harassment-and-hate-crimes-rise-as-coronavirus-spreads.

       *a man killed eight people:* Shannon Stapleton, "8 Dead in Atlanta Spa Shootings, with Fears of Anti-Asian Bias," *New York Times,* March 26, 2021, https://www.nytimes.com/live/2021/03/17/us/shooting-atlanta-acworth.

       *fewer than 500 incidents:* "White Supremacist Propaganda Spikes in 2020," Anti-Defamation League, https://www.adl.org/white-supremacist-propaganda-spikes-2020.

       *"garbage people":* "Antisemitism and the Radical Anti-Israel Movement on U.S. Campuses, 2019," Anti-Defamation League, https://www.adl.org/resources/reports/antisemitism-and-the-radical-anti-israel-movement-on-us-campuses-2019.

       *accused Israel of apartheid:* "Israel Practices 'Apartheid' — Rep. Betty McCollum," remarks by Rep. Betty McCollum at USCPR annual conference, U.S. Campaign for Palestinian Rights," September 29, 2018, https://uscpr.org/israel-practices-apartheid-rep-betty-mccollum/.

       *"slander":* Richard J. Goldstone, "Israel and the Apartheid Slander," *New York Times,* October 31, 2011, https://www.nytimes.com/2011/11/01/opinion/israel-and-the-apartheid-slander.html.

       *"Not One Dollar":* Martha Grevatt, "Commentary: Not One Dollar for Zionist Genocide!," *Workers World,* May 17, 2021, https://www.workers.org/2021/05/56475/.

50    *"We can learn about":* Jinnie Spiegler, interview with the author, March 8, 2021.

### 3. THE TOP OF THE PYRAMID

53    *Babylonian sacking of Jerusalem:* "Jews in Islamic Countries: Iran," Jewish Virtual Library, March 2021, https://www.jewishvirtuallibrary.org/jews-of-iran.

*most high-profile Jewish business figure:* David Green, "This Day in Jewish History / An Execution in Iran," *Haaretz,* May 9, 2014, https://www.haaretz.com/jewish/.premium-an-execution-in-iran-1.5247705.

*met with Ayatollah Khomeini:* "Iranian Jews Reassured in Talk with Khomeini," *New York Times,* May 16, 1979; Roya Hakakian, "How Iran Kept Its Jews," *Tablet,* December 30, 2014, https://www.tabletmag.com/sections/israel-middle-east/articles/how-iran-kept-its-jews.

*executing prominent members:* Karmel Melamed, "Iran's War Against the Jews Began in May 1979," *Times of Israel,* May 10, 2020.

*left the country:* Karmel Melamed, "The Iranian Revolution Was 40 Years Ago. Persian Jews in Los Angeles Are Still Feeling the Pain," Jewish Telegraphic Agency, February 25, 2019, https://www.jta.org/2019/02/25/global/the-iranian-revolution-was-40-years-ago-persian-jews-in-los-angeles-are-still-feeling-the-pain.

*Those who remained:* Apparently, some poorer Jews remained as well; see Larry Cohler-Esses, "How Iran's Jews Survive in Mullahs' World," World Jewish Congress, https://www.worldjewishcongress.org/en/about/communities/IR.

54    *the "hanging judge":* "'Hanging Judge' of Post-Shah Iran," *Irish Times,* December 6, 2003, https://www.irishtimes.com/news/hanging-judge-of-post-shah-iran-1.398107.

*Strict religious law:* Youssef M. Ibrahim, "Inside Iran's Cultural Revolution," *New York Times,* October 14, 1979, https://www.nytimes.com/1979/10/14/archives/inside-irans-cultural-revolution-iran.html.

*Universities were closed:* Farideh Farhi, "Cultural Policies in the Islamic Republic of Iran," Woodrow Wilson International Center for Scholars, November 2004, https://www.wilsoncenter.org/sites/default/files/media/documents/event/Farideh FarhiFinal.pdf; "The Stolen Revolution: Iranian Women of 1979," CBC Radio, March 8, 2019, https://www.cbc.ca/radio/ideas/the-stolen-revolution-iranian-women-of-1979-1.5048382.

56    *while hardly inevitable:* As one scholar observes, "The historical legacy of inter-community relations in Bosnia is mixed, but it doesn't lend itself to the easy conclusion that these groups were destined to fight each other in the way that they did"; see Sumantra Bose, "Bosnia: The Origins of the Conflict," London School of Economics and Political Science, November 16, 2000, http://fathom.lse.ac.uk/Features/122271/.

*populations acted opportunistically:* Adna Karamehic-Oates, interview with the author's research team, April 22, 2021.

*one another's religious holidays:* Karamehic-Oates, interview with the author's research team.

57    *"malign" opportunist:* Jones, *Genocide,* 318.

*"You should stay here":* Laura Silber and Allan Little, *Yugoslavia: Death of a Nation,* rev. ed. (New York: Penguin, 1997), 37–38.

*most ethnically mixed republic:* Paul Garde, "An Expert's Overview, Bosnia and Herzegovina, How They Became the Most Ethnically Diverse Republics in the Former Yugoslavia, and How the Different Ethnic Groups Interacted," *Frontline,* PBS, https://www.pbs.org/wgbh/pages/frontline/shows/karadzic/bosnia/bosnia.html.

58    *"a military campaign":* "Bosnia and Herzegovina, 1992–1995," United States Holocaust Memorial Museum, last updated July 2013, https://www.ushmm.org/genocide-prevention/countries/bosnia-herzegovina/case-study/background/1992-1995.

*"ethnically cleansed" villages:* Ann Petrila and Hasan Hasanovic, *Voices from Srebrenica: Survivor Narratives of the Bosnian Genocide* (Jefferson, NC: McFarland, 2021), 29–30.

*"Just like the Holocaust"*: Petrila and Hasanovic, *Voices from Srebrenica,* 33.

*"I didn't do anything wrong to anybody"*: Petrila and Hasanovic, *Voices from Srebrenica,* 37.

*"was extremely shocking"*: Karamehic-Oates, interview with the author's research team.

59    *"spoke of battles"*: Michael Sells, "Kosovo Mythology and the Bosnian Genocide," in Omer Bartov and Phyllis Mack, *In God's Name: Genocide and Religion in the Twentieth Century* (New York: Berghahn, 2001), 181.

*"lived on the blood"*: Sells, "Kosovo Mythology," 183.

*"an intense mass psychology"*: Sells, "Kosovo Mythology," 185.

*Muslims as evildoers:* Sells, "Kosovo Mythology," 188–90.

60    *"The message was"*: Karamehic-Oates, interview with the author's research team.

*"an almost unshakable image"*: Kent Fogg, "The Milošević Regime and the Manipulation of the Serbian Media," European Studies Conference, March 25, 2006.

*"a blinding fear"*: Refik Hodžić, "Dehumanisation of Muslims Made Karadzic an Icon of Far-Right Extremism," Justice Hub, March 22, 2019, https://justicehub.org/article/dehumanisation-muslims-made-karadzic-icon-far-right-extremism/.

61    *"stopped talking to me"*: Kenan Trebincevic and Susan Shapiro, *The Bosnia List: A Memoir of War, Exile, and Return* (New York: Penguin, 2014), 28.

*"It felt as if someone"*: Trebincevic and Shapiro, *The Bosnia List,* 39.

*"Soon Brcko will be rid"*: Trebincevic and Shapiro, *The Bosnia List,* 42, 44.

62    *the presence of "hate propaganda"*: Jones, *Genocide,* 567–71.

*"the Ten Stages of Genocide"*: Gregory Stanton, "The Ten Stages of Genocide," Genocide Watch, 1996, https://www.genocidewatch.com/tenstages. The ten stages, according to this model, are classification, symbolization, discrimination, dehumanization, organization, polarization, preparation, persecution, extermination, and denial.

63    *"a rhetorical and conceptual pillar"*: Jasmin Mujanović, "The Balkan Roots of the Far Right's 'Great Replacement' Theory," *Newlines,* March 12, 2021, https://newlines-mag.com/essays/the-balkan-roots-of-the-far-rights-great-replacement-theory/.

*listened to a Serb nationalist song:* Jovana Gec, "New Zealand Gunman Entranced with Ottoman Sites in Europe," *AP News,* March 16, 2019, https://apnews.com/article/49277b64c27541cbb7d7e9a07e32392d. See also Jasmin Mujanović, "Why Serb Nationalism Still Inspires Europe's Far Right," *Balkan Insight,* March 22, 2019, https://balkaninsight.com/2019/03/22/why-serb-nationalism-still-inspires-europes-far-right/.

*"From where I sit"*: Amra Sabic-El-Rayess, "Today's America Reminds Me of 1990s Bosnia and Herzegovina," *Aljazeera,* October 1, 2020, https://www.aljazeera.com/opinions/2020/10/1/from-90s-bosnia-to-modern-dayhistory-repeats-itself.

## 4. THE MAKING OF AN EXTREMIST

65    *"[the shooting made] it clear"*: Matt Stroud, "CEO of Surveillance Firm Banjo Once Helped KKK Leader Shoot Up a Synagogue," *OneZero,* April 28, 2020, https://onezero.medium.com/ceo-of-surveillance-firm-banjo-once-helped-kkk-leader-shoot-up-synagogue-fdba4ad32829. I base the narrative in this chapter on this article as well as on numerous conversations I've had with Damien Patton, former CEO of the software company Banjo. I also draw on an interview with Patton done by a team member of mine on March 6, 2021. Unless otherwise indicated, all quotations in this chapter come from that interview or my own conversations. When possible, I have

used that interview or my own conversations for the factual reconstructions. When Patton's account conflicted with the article's, I've opted to rely on Patton's reconstruction of events. I've used the article and its reporting to fill in gaps in Patton's memory and to establish facts I wasn't able to glean in my conversations or in the interview. I also drew on "KKK Leader Arrested on Shooting Charges," *Deseret News,* December 21, 1991, https://www.deseret.com/1991/12/21/18957920/kkk-leader-arrested-on-shooting-charges.

*pleaded guilty to the shooting:* "KKK Leader Pleads Guilty in Shooting," *Deseret News,* April 12, 1992, https://www.deseret.com/1992/4/12/18978168/kkk-leader-pleads-guilty-in-shooting; Stroud, "CEO of Surveillance Firm Banjo."

66 *"Jews . . . were felt to be":* Stroud, "CEO of Surveillance Firm Banjo."

67 *Some psychologists theorize:* Arie Kruglanski et al., "The Making of Violent Extremists," *Review of General Psychology* 22, no. 1, (2018): 107–12; Angela Fritz, "The Psychology of How Someone Becomes Radicalized," *Washington Post,* November 1, 2018, https://www.washingtonpost.com/science/2018/11/01/psychology-how-someone-becomes-radicalized/.

*their ties with social institutions:* Christoffer Carlsson et al., "A Life-Course Analysis of Engagement in Violent Extremist Groups," *British Journal of Criminology* 60, no. 1 (2020): 74–92.

*"feeling alone or lacking meaning":* Cathy Cassata, "Why Do People Become Extremists?," *Healthline,* September 18, 2017, https://www.healthline.com/health-news/why-do-people-become-extremists#Then-theres-the-lone-wolf.

*higher rates of childhood trauma:* "Trauma as a Precursor to Violent Extremism," National Consortium for the Study of Terrorism and Responses to Terrorism, April 2015, https://www.start.umd.edu/pubs/START_CSTAB_TraumaAsPrecursortoViolent Extremism_April2015.pdf.

*five types of people:* Tore Bjørgo and Hanna Munden, "What Explains Why People Join and Leave Far-Right Groups?," University of Oslo, last modified November 7, 2020, https://www.sv.uio.no/c-rex/english/groups/compendium/what-explains-why-people-join-and-leave-far-right-groups.html.

68 *classmate named Pablo:* At Patton's request, I've changed his classmate's name to protect his identity.

75 *"had a ticket":* Stroud, "CEO of Surveillance Firm Banjo." Patton recounted this event in court testimony reported in this article.

77 *resigned as Banjo's CEO:* Bubba Brown, "Banjo CEO Damien Patton Resigns in Wake of KKK Revelations," *Park Record,* May 11, 2020, https://www.parkrecord.com/news/banjo-ceo-damien-patton-resigns-in-wake-of-kkk-revelations/.

79 *"This was not something":* Stroud, "CEO of Surveillance Firm Banjo."

81 *Drop the ADL:* "Open Letter to Progressives: The ADL Is Not an Ally," #Drop TheADL, https://droptheadl.org/.

82 *In 2018, two men:* Matt Stevens, "Starbucks C.E.O. Apologizes After Arrests of 2 Black Men," *New York Times,* April 15, 2018, https://www.nytimes.com/2018/04/15/us/starbucks-philadelphia-black-men-arrest.html.

83 *extreme, hateful statements:* As one example among many, see Aaron Bandler, "Sarsour Says Israel 'Is Built on the Idea That Jews Are Supreme to Everyone Else,'" *Jewish Journal,* December 2, 2019, https://jewishjournal.com/news/united-states/307887/sarsour-says-israel-is-built-on-the-idea-that-jews-are-supreme-to-everyone-else/.

*"constantly attacking black and brown people":* "Women's March Leader Tamika Mallory At-

tacks Starbucks for Including ADL in Bias Trainings," Jewish Telegraphic Agency, April 18, 2018, https://www.jta.org/2018/04/18/united-states/womens-march-leader-tamika-mallory-attacks-starbucks-including-adl-bias-adviser.

*"an anti-Arab, anti-Palestinian organization":* Linda Sarsour, Facebook, April 17, 2018, https://www.facebook.com/linda.sarsour/posts/10156525223155572; Daniel J. Roth, "Women's March Leaders Slam ADL, Call Group 'Islamophobic,' Anti-Minority," *Jerusalem Post,* April 18, 2018, https://www.jpost.com/diaspora/womens-march-leaders-slam-adl-call-group-islamophobic-anti-minority-551213.

84  *wrote an essay:* Jonathan Greenblatt, "Why I'm Speaking to Students at J Street U," Medium, April 17, 2016, https://jonathan-g.medium.com/why-i-m-speaking-to-students-at-j-street-u-cd6499b6fd63.

   *released my remarks:* "Remarks by Jonathan A. Greenblatt," Anti-Defamation League, April 17, 2016, https://www.adl.org/news/article/remarks-by-jonathan-a-greenblatt #.VxOTHTArLIU.

   *outraged op-eds:* Isi Leibler, "Candidly Speaking: The ADL's Moral Equivalence: Israelis and Palestinians," *Jerusalem Post,* May 18, 2016, https://www.jpost.com/Opinion/CANDIDLY-SPEAKING-The-ADLs-moral-equivalence-Israelis-and-Palestinians-454327.

85  *Those on the Far Left claim:* Jacob Hutt and Alex Kane, "How the ADL's Israel Advocacy Undermines Its Civil Rights Work," *Jewish Currents,* February 8, 2021, https://jewishcurrents.org/how-the-adls-israel-advocacy-undermines-its-civil-rights-work/.

   *Detractors on the extreme right:* Seth Mandel, "The Shame of the Anti-Defamation League," *Commentary*, November 2018, https://www.commentarymagazine.com/articles/seth-mandel/the-shame-of-the-anti-defamation-league/.

86  *"destroying the ADL":* Alex VanNess, "Jonathan Greenblatt Is Destroying the Anti-Defamation League," *New York Post,* December 9, 2016, https://nypost.com/2016/12/09/jonathan-greenblatt-is-destroying-the-anti-defamation-league/; Charles Jacobs and Avi Goldwasser, "America's Failed Jewish Leadership Must Resign," *Algemeiner,* December 23, 2019, https://www.algemeiner.com/2019/12/23/americas-failed-jewish-leadership-must-resign/. Far Right voices publicly bemoaned my appointment as ADL's new head even before I started on the job because I was recruited from the Obama White House, which they deemed anti-Israel on account of the administration's various policy positions.

87  *"welcoming Zionists to campus":* Daniel J. Roth, "Arab Prof. Under Fire as President's Apology Turns to 'Declaration of War,'" *Jerusalem Post,* March 22, 2018, https://www.jpost.com/middle-east/dozens-of-organizations-condemn-prof-for-anti-zionists-facebook-posts-546774.

   *indigenous Uyghur population:* See, for instance, "China: Crimes Against Humanity in Xinjiang," Human Rights Watch, April 19, 2021, https://www.hrw.org/news/2021/04/19/china-crimes-against-humanity-xinjiang.

88  *"weary of fellow Muslims":* @Zahra Billoo, Twitter, September 10, 2018, https://twitter.com/ZahraBilloo/status/1039264852588552192.

   *"complicit in the prison-industrial complex":* "Antisemitism and the Radical Anti-Israel Movement on U.S. Campuses, 2019," Anti-Defamation League, https://www.adl.org/resources/reports/antisemitism-and-the-radical-anti-israel-movement-on-us-campuses-2019#antisemitic-themes.

   *there is no single path:* As one expert puts it, "The many models to date show there is no one profile and no one pathway that people take. Existing research shows that different pathways lead to radicalisation, while different people on a shared pathway either radicalise or do not"; see Julianna Photopoulos, "Why Don't Most People Become Radicalised?," *Horizon*, June 14, 2018, https://horizon-magazine.eu/article/why-don-t-most-people-become-radicalised.html.

89    *aren't dangerous psychopaths:* Intriguing research suggests that some aspects of our psychological makeup might incline us more or less toward extremism. As one researcher put it: "Subtle difficulties with complex mental processing may subconsciously push people towards extreme doctrines that provide clearer, more defined explanations of the world, making them susceptible to toxic forms of dogmatic and authoritarian ideologies"; see Duaa Nasir, "What Kinds of People Become Extremists?," *Medium*, March 1, 2021, https://issuu.com/mediumutm/docs/2021-03-01_issue_19/12.

*"When I ask them why":* Peter Byrne, "Anatomy of Terror: What Makes Normal People Become Extremists?," *New Scientist*, August 16, 2017, https://www.newscientist.com/article/mg23531390-700-anatomy-of-terror-what-makes-normal-people-become-extremists/.

90    *"The lowest common denominator":* Madeline Roache, "This Researcher Juggled Five Different Identities to Go Undercover with Far-Right and Islamist Extremists. Here's What She Found," *Time*, February 18, 2020, https://time.com/5785779/extremist-networks-julia-ebner/.

*someone fired:* Jordan Buie and Stacey Barchenger, "Institutions on Alert After Shot Fired at West End Synagogue," *Tennessean*, April 13, 2015, https://www.tennessean.com/story/news/crime/2015/04/13/shots-fired-at-west-end-synagogue/25711391/.

# 5. HATE BOOSTERS

93    *"They were playing films of the Nazis":* Jacob Judah, "Shiva 'Zoom-Bombed' by Neo-Nazis Showing Films of Hitler and the Holocaust," *JC*, June 14, 2021, https://www.thejc.com/news/uk/shiva-zoom-bombed-by-neo-nazis-showing-films-of-hitler-and-the-holocaust-1.505626; Mathilde Frot, "Virtual Shiva Allegedly Zoom-Bombed with Videos of Adolf Hitler," *Jewish News*, August 10, 2020, https://jewishnews.timesofisrael.com/virtual-shiva-allegedly-zoom-bombed-with-images-of-adolf-hitler/. I tweeted about this incident at the time; see Aaron Bandler, "Virtual Shivah in UK Zoombombed with Swastikas, Hitler Clips," *Jewish Journal*, August 11, 2020, https://jewishjournal.com/featured/320153/virtual-shivah-zoombombed-with-swastikas-hitler-clips/.

*numerous such events:* "What Is 'Zoombombing' and Who Is Behind It?," Anti-Defamation League, May 4, 2020, https://www.adl.org/blog/what-is-zoombombing-and-who-is-behind-it. In March 2020, the FBI issued a formal warning about Zoom-bombing; see Kristen Setera, "FBI Warns of Teleconferencing and Online Classroom Hijacking During COVID-19 Pandemic," FBI Boston, March 30, 2020, https://www.fbi.gov/contact-us/field-offices/boston/news/press-releases/fbi-warns-of-teleconferencing-and-online-classroom-hijacking-during-covid-19-pandemic.

*"Jews in the ovens":* Cnaan Liphshiz, "Zoom Bombers Shout 'Jews in the Ovens' at Online Holocaust Book Launch in Italy," *Times of Israel*, January 18, 2021, https://www.timesofisrael.com/zoom-bombers-shout-jews-in-the-ovens-at-online-holocaust-book-launch-in-italy/; Cody Levine, "Virtual Tisha Be'av Service Hit by An-

tisemitic Hate, Calls to Bomb Israel," *Jerusalem Post,* August 2, 2020, https://www.jpost.com/diaspora/virtual-tisha-bav-service-hit-by-antisemitic-hate-calls-to-bomb-israel-637079.

94    *over two-fifths of Americans:* "Online Hate and Harassment: The American Experience 2021," Anti-Defamation League, https://www.adl.org/online-hate-2021#executive-summary.

95    *"an alt-right rabbit hole":* Kevin Roose, "The Making of a YouTube Radical," *New York Times,* June 8, 2019, https://www.nytimes.com/interactive/2019/06/08/technology/youtube-radical.html.

       *implicated social media:* Daniela Hernandez and Parmy Olson, "Isolation and Social Media Combine to Radicalize Violent Offenders," *Wall Street Journal,* August 5, 2019, https://www.wsj.com/articles/isolation-and-social-media-combine-to-radicalize-violent-offenders-11565041473.

       *slaughtered nine African-Americans:* Mark Berman, "Prosecutors Say Dylann Roof 'Self-Radicalized' Online, Wrote Another Manifesto in Jail," *Washington Post,* August 22, 2016, https://www.washingtonpost.com/news/post-nation/wp/2016/08/22/prosecutors-say-accused-charleston-church-gunman-self-radicalized-online/.

96    *the most violent antisemitic attack in American history:* "Gab and 8chan: Home to Terrorist Plots Hiding in Plain Sight," Anti-Defamation League, https://www.adl.org/resources/reports/gab-and-8chan-home-to-terrorist-plots-hiding-in-plain-sight; Kevin Roose, "On Gab, an Extremist-Friendly Site, Pittsburgh Shooting Suspect Aired His Hatred in Full," *New York Times,* October 28, 2018, https://www.nytimes.com/2018/10/28/us/gab-robert-bowers-pittsburgh-synagogue-shootings.html.

       *90 percent of "lone actors":* Hernandez and Olson, "Isolation and Social Media Combine."

       *lone wolves:* Farah Pandith, "Teen Terrorism Inspired by Social Media Is on the Rise. Here's What We Need to Do," NBC News, March 22, 2021, https://www.nbcnews.com/think/opinion/teen-terrorism-inspired-social-media-rise-here-s-what-we-ncna1261307.

       *hate groups like Feuerkrieg Division:* Michael Kunzelman and Jari Tanner, "He Led a Neo-Nazi Group Linked to Bomb Plots. He was 13," ABC News, April 11, 2020, https://abcnews.go.com/US/wireStory/led-neo-nazi-group-linked-bomb-plots-13-70099974hate grou.

       *ADL researchers analyzed tens of thousands of Facebook posts:* "Computational Propaganda and the 2020 U.S. Presidential Election: Antisemitic and Anti-Black Content on Facebook and Telegram," Anti-Defamation League, https://www.adl.org/resources/reports/computational-propaganda-and-the-2020-election#bibliography.

       *"online propaganda can feed acts of violent terror":* "Through Conspiracies and Coded Language, White Supremacists Use Social Media Networks to Aid and Abet Terror, New Study Reveals," Anti-Defamation League, April 9, 2019, https://www.adl.org/news/press-releases/through-conspiracies-and-coded-language-white-supremacists-use-social-media.

97    *the world's oldest hatred:* Frank Bruni, "The Oldest Hatred, Forever Young," *New York Times,* April 14, 2014, https://www.nytimes.com/2014/04/15/opinion/the-oldest-hatred-forever-young.html?ref=frankbruni&_r=0.

       *"the most durable and pliable of all conspiracy theories":* Richard Cohen, "Anti-Semitism Is Once Again on the March in Europe," *Washington Post,* February 2, 2015, https://www.washingtonpost.com/opinions/richard-cohen-anti-semitism-is-once-again-on-the-march-in-europe/2015/02/02/9520ae7a-ab0d-11e4-abe8-e1ef60ca26de_story.html.

*dating from the early twentieth century:* Stephen Whitfield, "Why the 'Protocols of the Elders of Zion' Is Still Pushed by Anti-Semites," *Brandeis NOW,* September 2, 2020, https://www.brandeis.edu/now/2020/september/whitfield-conversation.html.

*so-called Boston Aristocracy:* Mark Cheathem, "Conspiracy Theories Abounded in 19th-Century American Politics," *Smithsonian,* April 11, 2019, https://www.smithsonianmag.com/history/conspiracy-theories-abounded-19th-century-american-politics-180971940/.

*where they are rampant:* See, for instance, Chris Mills Rodrigo, "Conspiracy Theories Run Rampant Online Amid Floyd Protests," *Hill,* June 3, 2020, https://thehill.com/policy/technology/500817-conspiracy-theories-run-rampant-online-amid-floyd-protests. One 2020 survey found that almost two-thirds of Americans felt that social media had a negative impact, with almost 30 percent of those citing misinformation as a problem; see Brooke Auxier, "64% of Americans Say Social Media Have a Mostly Negative Effect on the Way Things Are Going in the U.S. Today," Pew Research Center, October 15, 2020, https://www.pewresearch.org/fact-tank/2020/10/15/64-of-americans-say-social-media-have-a-mostly-negative-effect-on-the-way-things-are-going-in-the-u-s-today/.

*most obvious example is QAnon:* "QAnon," Anti-Defamation League, https://www.adl.org/qanon.

*"a rebranded version":* Gregory Stanton, "QAnon Is a Nazi Cult, Rebranded," *Just Security,* September 9, 2020, https://d0dbb2cb-698c-4513-aa47-eba3a335e06f.filesusr.com/ugd/137a5c_a320e0412b194d18b6898a144b531a56.pdf.

*classic antisemitic notion:* "Blood Libel: A False, Incendiary Claim Against Jews," Anti-Defamation League, https://www.adl.org/education/resources/glossary-terms/blood-libel.

*to achieve world domination:* A range of conspiracy theories, including antisemitic ones, circulated online about the COVID vaccines; see "Racist, Extremist, Antisemitic Conspiracies Surround Coronavirus Vaccine Rollout," Anti-Defamation League, December 29, 2020, https://www.adl.org/blog/racist-extremist-antisemitic-conspiracies-surround-coronavirus-vaccine-rollout.

*"'global elites' will use the pandemic":* "'The Great Reset' Conspiracy Flourishes Amid Continued Pandemic," Anti-Defamation League, December 29, 2020, https://www.adl.org/blog/the-great-reset-conspiracy-flourishes-amid-continued-pandemic.

98 *fail to act:* According to one report, "One investigation found that copies of QAnon videos removed from Facebook and Twitter remained in circulation on those same platforms, and a cross-platform study of items of COVID-19 related content identified as misinformation by fact-checking organizations found that no action was taken with regard to 59% of such items on Twitter, 27% of such items on YouTube and 24% of such items on Facebook, despite platform policies which suggested that action would be taken promptly"; see Alexis Weger, "Social Media Helps Perpetuate Conspiracy Theories That Can Lead to Radicalization," *Homeland Security Today,* February 16, 2021, https://www.hstoday.us/subject-matter-areas/information-technology/social-media-helps-perpetuate-conspiracy-theories-that-can-lead-to-radicalization/.

*peddled conspiracy theories:* Debra Saunders, "Ron Brown's Body Lies A-Moldering," *SFGATE,* February 14, 2012, https://www.sfgate.com/opinion/saunders/article/Ron-Brown-s-Body-Lies-A-Moldering-3328675.php; "Does Bill Clinton Run Murder, Inc.?," *Slate,* February 18, 1999, https://slate.com/news-and-politics/1999/02/does-bill-clinton-run-murder-inc.html; Terry Frieden, "Justice Concludes No Evidence of Crime in Ron Brown Death," CNN, January 8, 1998, https://edition.cnn.com/ALLPOLITICS/1998/01/08/justice.brown/.

99   *"makes them feel safer"*: Anne Applebaum, *Twilight of Democracy: The Seductive Lure of Authoritarianism* (New York: Doubleday, 2020), 106, 113.

*"Post-truth is pre-fascism"*: Timothy Snyder, *On Tyranny: Twenty Lessons from the Twentieth Century* (New York: Tim Duggan Books, 2017), 71.

*Twitter permanently barred the conspiracy theorist Alex Jones:* Phil McCausland and Ben Collins, "Twitter Permanently Bans Conspiracy Theorist Alex Jones and Website Infowars," NBC News, September 6, 2018, https://www.nbcnews.com/tech/tech-news/twitter-permanently-bans-conspiracy-theorist-alex-jones-website-infowars-n907261.

*posts containing white-nationalist and white-supremacist messages:* David Ingram and Ben Collins, "Facebook Bans White Nationalism from Platform After Pressure from Civil Rights Groups," NBC News, March 27, 2019, https://www.nbcnews.com/tech/tech-news/facebook-bans-white-nationalism-after-pressure-civil-rights-groups-n987991. Although frequently conflated and overlapping, white nationalism and white supremacism are not the same. White supremacists believe in the inherent superiority of the white race, while white nationalists seek to forge a nation-state around a shared ethnic or cultural heritage associated with "whiteness" (for instance, an Anglo-Saxon state). In practice, many white nationalists are also white supremacists.

*independent panel of experts:* Kate Klonick, "Inside the Making of Facebook's Supreme Court," *New Yorker,* February 12, 2021, https://www.newyorker.com/tech/annals-of-technology/inside-the-making-of-facebooks-supreme-court; Emily Bell, "Facebook Has Beefed Up Its 'Oversight Board,' but Any New Powers Are Illusory," *Guardian,* April 14, 2021, https://www.theguardian.com/commentisfree/2021/apr/14/facebook-has-beefed-up-its-oversight-board-but-any-new-powers-are-illusory.

100   *Stop Hate for Profit website:* See https://www.stophateforprofit.org/productrecommendations.

*hit the Pause button:* "ADL, NAACP, Sleeping Giants, Common Sense, Free Press and Color of Change Call for Global Corporations to Pause Advertising on Facebook to Stop Hate Online," Anti-Defamation League, June 17, 2020, https://www.adl.org/news/press-releases/adl-naacp-sleeping-giants-common-sense-free-press-and-color-of-change-call-for.

101   *Holocaust denialism as hate speech:* Jessica Guynn, "Facebook Still Has Holocaust Denial Content Three Months After Mark Zuckerberg Pledged to Remove It," *USA Today,* January 27, 2021, https://www.usatoday.com/story/tech/2021/01/27/facebook-holocaust-denial-zuckerberg-twitter-youtube-twitch-reddit/4269035001/.

*hired a vice president for civil rights:* Megan Rose Dickey, "Facebook Hires a VP of Civil Rights," *TechCrunch,* January 11, 2021, https://techcrunch.com/2021/01/11/facebook-hires-a-vp-of-civil-rights.

*"very rough":* Megan Graham, "Facebook's Ad Chief Explains How the Boycotts Changed the Company," CNBC, October 22, 2020, https://www.cnbc.com/2020/10/22/facebooks-ad-chief-talks-ad-boycotts-.html.

102   *cracked down on links:* "Online Hate and Harassment: The American Experience 2021," Anti-Defamation League, https://www.adl.org/online-hate-2021#introduction.

*ban Trump permanently:* Carla Jimenez, "Far-Right Extremists on Social Media Aren't Going Away — They're Hunkering Down," *CPR News,* January 20, 2021, https://

www.cpr.org/2021/01/20/far-right-extremists-on-social-media-arent-going-away-theyre-hunkering-down/; Shannon Bond, "Unwelcome on Facebook and Twitter, QAnon Followers Flock to Fringe Sites," NPR, January 31, 2021, https://www.npr.org/2021/01/31/962104747/unwelcome-on-facebook-twitter-qanon-followers-flock-to-fringe-sites; Julie Jammot, "Extremists Turn Elsewhere After Social Media Giants' Crackdown," *Times of Israel,* February 23, 2021, https://www.timesofisrael.com/extremists-turn-elsewhere-after-social-media-giants-crackdown/; Martin Pengelly, "Trump Will Use 'His Own Platform' to Return to Social Media After Twitter Ban," *Guardian,* March 21, 2021, https://www.theguardian.com/us-news/2021/mar/21/trump-twitter-ban-social-media-own-platform.

103 *tweets out blatantly antisemitic messages:* David Weinberg, "Twitter Must De-Platform Iran's Supreme Leader," Anti-Defamation League, March 19, 2021, https://www.adl.org/blog/twitter-must-de-platform-irans-supreme-leader.

*Young Pharoah:* "Antisemitic Rapper 'Young Pharoah' Spews Bigotry, Racist Hate," Anti-Defamation League, February 26, 2021, https://www.adl.org/blog/antisemitic-rapper-young-pharoah-spews-bigotry-racist-hate.

*Holocaust-denial-management report card:* "Online Holocaust Denial Report Card: An Investigation of Online Platforms' Policies and Enforcement," Anti-Defamation League, https://www.adl.org/holocaust-denial-report-card#the-online-holocaust-denial-report-card-explained-.

*unchanged from 2020:* "Online Hate and Harassment: The American Experience 2021," Anti-Defamation League, https://www.adl.org/online-hate-2021.

*"creeping along toward firmer action":* Barbara Ortutay, "Facebook Forced to Reckon with Misinformation, Hate Speech," *Chicago Tribune,* March 24, 2021, https://www.chicagotribune.com/business/ct-biz-facebook-speech-crackdown-20210324-mkjy5uuxnbbxfplanq3bkqm66y-story.html.

104 *thirty-one million bottles:* Howard Markel, "How the Tylenol Murders of 1982 Changed the Way We Consume Medication," *PBS NewsHour,* September 29, 2014, https://www.pbs.org/newshour/health/tylenol-murders-1982.

*spending more than one hundred million dollars:* Judith Rehak, "Tylenol Made a Hero of Johnson & Johnson: The Recall That Started Them All," *New York Times,* March 23, 2002, https://www.nytimes.com/2002/03/23/your-money/IHT-tylenol-made-a-hero-of-johnson-johnson-the-recall-that-started.html.

105 *basic and practicable steps:* For a brief synopsis of what social media companies should do to combat hate, please see the "Recommendations" section of "Online Hate and Harassment: The American Experience 2021," Anti-Defamation League, https://www.adl.org/online-hate-2021#recommendations.

*"weary giants of flesh and steel":* John Perry Barlow, "A Declaration of the Independence of Cyberspace," February 8, 1996, https://www.eff.org/cyberspace-independence.

106 *"There has always been a fringe element":* Huffman made this comment on December 12, 2020, at a Zoom meeting of the advisory board of the ADL Center for Technology and Society.

107 *sued Fox News for $1.6 billion:* Merrit Kennedy and Bill Chappell, "Dominion Voting Systems Files $1.6 Billion Defamation Lawsuit Against Fox News," NPR, March 26, 2021, https://www.npr.org/2021/03/26/981515184/dominion-voting-systems-files-1-6-billion-defamation-lawsuit-against-fox-news.

*filed a similar lawsuit:* Jonah Bromwich and Ben Smith, "Fox News Is Sued by Election Technology Company for Over $2.7 Billion," *New York Times,* April 27, 2021, https://www.nytimes.com/2021/02/04/business/media/smartmatic-fox-news-lawsuit.html.

108  *allows internet companies to host user content:* Daisuke Wakabayashi, "Legal Shield for Social Media Is Targeted by Lawmakers," *New York Times,* May 28, 2020, https://www.ny times.com/2020/05/28/business/section-230-internet-speech.html.

*essential safeguard of freedom of expression:* "Section 230 of the Communications Decency Act," Electronic Frontier Foundation, accessed June 18, 2021, https://www.eff.org/issues/cda230.

*compensate them for any harm:* In some cases, the harm is considerable; see Kashmir Hill, "A Vast Web of Vengeance," *New York Times,* February 2, 2021, https://www.nytimes.com/2021/01/30/technology/change-my-google-results.html.

109  *extreme, provocative, angry:* Jamie Seidel, "How Facebook, Google Algorithms Feed on Hate Speech, Rage," *New Zealand Herald,* September 5, 2020, https://www.nzherald.co.nz/business/how-facebook-google-algorithms-feed-on-hate-speech-rage/W7LP GNG6SKGN6Q6FN2O3HW6WVM/; Katherine J. Wu, "Radical Ideas Spread Through Social Media. Are the Algorithms to Blame?," *Nova,* March 28, 2019, https://www.pbs.org/wgbh/nova/article/radical-ideas-social-media-algorithms/; "Social Media Platforms Face Reckoning Over Hate Speech," *VOA News,* June 30, 2020, https://www.voanews.com/silicon-valley-technology/social-media-platforms-face-reckoning-over-hate-speech; Jack Nicas, "How YouTube Drives People to the Internet's Darkest Corners," *Wall Street Journal,* February 7, 2018, https://www.wsj.com/articles/how-youtube-drives-viewers-to-the-internets-darkest-corners-1518020478?mod=article_inline.

*"There's a spectrum on YouTube":* Roose, "The Making of a YouTube Radical."

110  *"Our algorithms exploit":* Jeff Horwitz and Deepa Seetharaman, "Facebook Executives Shut Down Efforts to Make the Site Less Divisive," *Wall Street Journal,* May 26, 2020, https://www.wsj.com/articles/facebook-knows-it-encourages-division-top-execu tives-nixed-solutions-11590507499.

*"YouTube [has] purposefully ignored warnings":* Luke Munn, "Angry by Design: Toxic Communication and Technical Architectures," *Humanities and Social Sciences Communications* 7 (July 30, 2020), https://www.nature.com/articles/s41599-020-00550-7.

*"always [does] just enough":* Karen Hao, "How Facebook Got Addicted to Spreading Misinformation," *MIT Technology Review,* March 11, 2021, https://www.technology review.com/2021/03/11/1020600/facebook-responsible-ai-misinformation/.

*killing people:* Salvador Rodriguez, "Biden on Facebook: 'They're Killing People' with Vaccine Misinformation," CNBC, July 16, 2021, https://www.cnbc.com/2021/07/16/white-house-says-facebook-needs-to-do-more-to-fight-vaccine-misinformation.html.

111  *profits of $32 billion:* "Facebook Reports Fourth Quarter and Full Year 2020 Results," Facebook, January 27, 2021, https://investor.fb.com/investor-news/press-release-details/2021/Facebook-Reports-Fourth-Quarter-and-Full-Year-2020-Results/default.aspx.

112  *61 percent of respondents:* "American Attitudes Toward Extremist Threats: A Survey Following the Events at the U.S. Capitol," Anti-Defamation League, 2021, https://www.adl.org/american-attitudes-toward-extremist-threats.

*killed thousands of Rohingya Muslims:* "Myanmar Rohingya: What You Need to Know

About the Crisis," BBC News, January 23, 2020, https://www.bbc.com/news/world-asia-41566561; https://www.cfr.org/backgrounder/rohingya-crisis.

*"for months, if not years":* Mathew Ingram, "Facebook Slammed by UN for Its Role in Myanmar Genocide," *Columbia Journalism Review,* November 8, 2018, https://www.cjr.org/the_media_today/facebook-un-myanmar-genocide.php.

*"substantively contributed":* Tom Miles, "U.N. Investigators Cite Facebook Role in Myanmar Crisis," Reuters, March 12, 2018, https://www.reuters.com/article/us-myanmar-rohingya-facebook/u-n-investigators-cite-facebook-role-in-myanmar-crisis-idUSKCN1GO2PN.

*"do more to ensure":* Alex Warofka, "An Independent Assessment of the Human Rights Impact of Facebook in Myanmar," Facebook, November 5, 2018, https://about.fb.com/news/2018/11/myanmar-hria/.

## 6. AMERICAN BERSERK

114   *"The excitement was intense":* "Awful Murder, and Savage Barbarity," *St. Louis Observer,* May 5, 1835, http://civilwarmo.org/gallery/item/CWMO-61?nojs=1.

      *the first lynching:* Walter Johnson and Jamala Rogers, "No Excuse for Forgetting Black St. Louis," *St. Louis Post-Dispatch,* July 10, 2018, https://scholar.harvard.edu/files/wjohnson/files/johnson_and_rogers-no_excuse_for_forgetting_black_st_louis.pdf; Russell Contreras, "Vile US History of Lynching of People of Color," Associated Press, June 15, 2020, https://www.kpbs.org/news/2020/jun/15/ap-explains-vile-us-history-of-lynching-of-people/.

115   *"cannot come from abroad":* Abraham Lincoln, "The Perpetuation of Our Political Institutions: Address Before the Young Men's Lyceum of Springfield, Illinois," Abraham Lincoln Online, January 27, 1838, http://www.abrahamlincolnonline.org/lincoln/speeches/lyceum.htm.

      *seem especially prescient:* Bret Stephens, "Lincoln Knew in 1838 What 2021 Would Bring," *New York Times,* January 18, 2021, https://www.nytimes.com/2021/01/18/opinion/trump-lincoln-mobs-democracy.html.

      *"is not only native":* Judith Shulevitz, "The Indigenous American Berserk," *Slate,* October 14, 2004, https://slate.com/news-and-politics/2004/10/the-indigenous-american-berserk.html. The phrase appeared in *American Pastoral:* "The daughter who transports him out of the longed-for American pastoral and into everything that is its antithesis and its enemy, into the fury, the violence, and the desperation of the counter-pastoral—into the indigenous American berserk"; see Philip Roth, *American Pastoral* (New York: Vintage, 1997), 86.

116   *with a global population:* There were about 14.7 million Jews in 2020 and about 15 million Ismailis as of 2015 see "Karachi Bus Massacre: Who Are the Ismailis?," BBC, May 13, 2015, https://www.bbc.com/news/world-asia-32721136#:~:text=How%20many%20Ismailis%20are%20there,in%20India%2C%20Afghanistan%20and%20Africa; "Global Jewish Population Hits 14.7 Million," *Australian Jewish News,* April 21, 2020, https://ajn.timesofisrael.com/global-jewish-population-hits-14-7-million/.

117   *"fake Jew Zionists":* Elizabeth Glass for Senate NY 27, Facebook, May 31, 2021, https://www.facebook.com/GlassforSovereignty/posts/332684781623547.

118   *during the writing of this book:* Melissa Koenig, "'Allahu Akhbar!': Pro-Palestinian Protest-

ers Chant as They Burn Israeli Flag in March Through NYC That Left Diners Shocked and Saw 'Blood' Thrown at BlackRock Offices," *Daily Mail,* June 11, 2021, https://www.dailymail.co.uk/news/article-9678709/Pro-Palestinian-group-BURNS-Israeli-flag-protest-businesses-ties-Israel.html.

*removing or otherwise vandalizing:* See, for instance, Pat Thomas, "Israeli and Virginia Tech Flags Stolen from Campus," WDBJ7, May 17, 2021, https://www.wdbj7.com/2021/05/17/israeli-and-virginia-tech-flags-stolen-from-campus/.

*incidents spiking 115 percent:* Emily Shapiro, "Antisemitism Surged Across US During Gaza Conflict, Part of Multi-Year Rise: Advocates," ABC News, June 10, 2021, https://abcnews.go.com/US/antisemitism-surged-us-gaza-conflict-part-multi-year/story?id=78092408.

*more than 500 percent rise:* Hannah Gal, "A Spike in Antisemitism Has British Jewry Worrying for the Future," *Jerusalem Post,* June 10, 2021, https://www.jpost.com/diaspora/antisemitism/a-spike-in-antisemitism-has-british-jewry-worrying-for-the-future-670709.

119   *a 1 percent chance:* "Risk Analysis for United States of America," Early Warning Project, United States Holocaust Memorial Museum, December 10, 2020, https://earlywarningproject.ushmm.org/countries/united-states-of-america#analysis.

   *a ten-stage process:* Stanton, "The Ten Stages of Genocide.".

120   *evidence for six of the ten stages:* Gregory Stanton, "United States of America," Genocide Watch, https://www.genocidewatch.com/united-states-of-america.

   *"I see white-supremacist groups":* Gregory Stanton, interview with the author's research team, April 27, 2021.

121   *"I don't think it's at all impossible":* Other genocide experts have also sounded warnings in recent years; see David Brennan, "U.S. Showing 'Many' Genocide Warning Signs Under Trump, Expert Says: 'I Am Very, Very Worried,'" *Newsweek,* January 24, 2020, https://www.newsweek.com/us-showing-many-genocide-warning-signs-donald-trump-expert-very-worried-1483817.

   *the overheated rhetoric:* Caroline Kelly, "Ocasio-Cortez on Calling Detention Centers 'Concentration Camps': We Have to 'Learn from Our History,'" CNN, June 27, 2019, https://www.cnn.com/2019/06/27/politics/alexandria-ocasio-cortez-concentration-camps-the-lead-cnntv/index.html.

   *Yale professor Tim Snyder:* Timothy Snyder, interview with the author, April 15, 2021.

122   *"We've entered a world":* Steven Levitsky, interview with the author's research team, April 29, 2021.

123   *Other experts worry:* Some have warned that the United States is growing closer to insta-bility and civil conflict; see Matthew Gault, "Is the U.S. Already in a New Civil War?," *Vice,* October 27, 2020, https://www.vice.com/en/article/qjp48x/is-the-us-already-in-a-new-civil-war; Monica Duffy Toft, "How Civil Wars Start," *Foreign Policy,* February 18, 2021, https://foreignpolicy.com/2021/02/18/how-civil-wars-start/. Some are more sanguine; see Richard Hanania, "Americans Hate Each Other. But We Aren't Headed for Civil War," *Washington Post,* October 29, 2020, https://www.washington-post.com/outlook/civil-war-united-states-unlikely-violence/2020/10/29/3a143936-0f0f-11eb-8074-0e943a91bf08_story.html. Of the chances of civil war, one expert has this to say: "How great is the risk of America going completely off the rails in the short to medium term? I wouldn't want to bet big money on it. But the odds are improv-ing all the time" (Damon Linker, "The Worst-Case Scenario for America's Immediate

Future," *Week,* January 15, 2021, https://theweek.com/articles/960957/worstcase-scenario-americas-immediate-future.

*mounted an insurgency:* John Dorney, "The Northern Ireland Conflict 1968–1998—an Overview," *Irish Story,* February 9, 2015, https://www.theirishstory.com/2015/02/09/the-northern-ireland-conflict-1968-1998-an-overview/#.YMjwbjZKjyI.

*"You're going to have a bombing here":* Barbara F. Walter, interview with author's research team, May 10, 2021. Walter made some of the arguments cited in this chapter in an interview given just after the January 6, 2021, insurrection; see Barbara Walter, "Political Scientist Warns of a Second Civil War After Capitol Riot," KPBS, January 7, 2021, https://www.kpbs.org/podcasts/kpbs-midday-edition-segments/2021/jan/07/political-scientist-warns-second-civil-war-after-c/.

## 7. FIGHTING HATE IN EVERYDAY LIFE

129   *"white genocide":* "White Genocide," Anti-Defamation League, 2021, https://www.adl.org/resources/glossary-terms/white-genocide.

*"Jews will not replace us":* "'The Great Replacement': An Explainer," Anti-Defamation League, 2021, https://www.adl.org/resources/backgrounders/the-great-replacement-an-explainer; https://www.adl.org/education/references/hate-symbols/you-will-not-replace-us.

*publicly espousing xenophobic and anti-immigrant sentiments:* Jonathan Greenblatt, "ADL Letter to Fox News Condemns Tucker Carlson's Impassioned Defense of 'Great Replacement Theory,'" Anti-Defamation League, April 9, 2021, https://www.adl.org/news/media-watch/adl-letter-to-fox-news-condemns-tucker-carlsons-impassioned-defense-of-great.

*"I know that the left":* Chris Cillizza, "How the Ugly, Racist White 'Replacement Theory' Came to Congress," CNN, April 15, 2021, https://www.cnn.com/2021/04/15/politics/scott-perry-white-replacement-theory-tucker-carlson-fox-news/index.html.

130   *3.4 million nightly viewers:* "Fox News Channel Reclaims Lead Sweeping Total Day and Primetime Viewers and Demo for the Month of March," *Business Wire,* March 30, 2021, https://www.businesswire.com/news/home/20210330005946/en/FOX-News-Channel-Reclaims-Lead-Sweeping-Total-Day-and-Primetime-Viewers-and-Demo-for-the-Month-of-March.

*wrote an open letter:* Greenblatt, "ADL Letter to Fox News."

131   *"comments on immigration":* David Sacks, "Get Ready for the 'No-Buy' List," *Common Sense with Bari Weiss,* July 30, 2021, https://bariweiss.substack.com/p/get-ready-for-the-no-buy-list.

*wrote back defending Carlson:* Thomas Moore, "Lachlan Murdoch Responds to Call for Tucker Carlson's Firing," *Hill,* April 13, 2021, https://thehill.com/homenews/media/547854-lachlan-murdoch-responds-to-call-for-tucker-carlsons-firing?rl=1.

*letter responding to Murdoch:* Jonathan Greenblatt, "ADL Response to Lachlan Murdoch, Chairman and CEO of Fox News, on Tucker Carlson," Anti-Defamation League, April 12, 2021, https://www.adl.org/news/media-watch/adl-response-to-lachlan-murdoch-chairman-and-ceo-of-fox-news-on-tucker-carlson.

*called on brands:* Lee Moran, "Anti-Defamation League Slams Tucker Carlson, Tells Ad-

vertisers to 'Choose a Side,'" *HuffPost,* April 21, 2021, https://www.huffpost.com/
entry/adl-tucker-carlson-advertisers_n_607fe040e4b047b9f8b46f15.

*"Commit to this fight":* "Jonathan Greenblatt's Remarks to the World Federation of Advertisers," Anti-Defamation League, April 20, 2021, https://www.adl.org/news/article
/jonathan-greenblatts-remarks-to-the-world-federation-of-advertisers.

133  *four strategies to try:* This material originally appeared in the ADL's "Antisemitism Uncovered Toolkit: Resources to Stand Up, Share Facts, and Show Strength Against Hate,"
Anti-Defamation League, 2021, https://www.adl.org/antisemitism-uncovered-
toolkit-resources-to-stand-up-share-facts-and-show-strength-against-hate.

134  *some questions to consider:* This material is reproduced and partially adapted from the
ADL handout "Taking a Stand: A Student's Guide to Stopping Name-Calling and
Bullying," Anti-Defamation League, 2005, https://www.adl.org/sites/default/files/
documents/assets/pdf/education-outreach/Taking-a-Stand-color.pdf.

136  *encounter hate in public places:* Material in this and the following paragraphs reproduced
largely verbatim from "Antisemitism Uncovered Toolkit."

139  *how to distinguish between the two:* As we at the ADL teach, criticizing Israel isn't inherently antisemitic, but it crosses the line when it delegitimizes or denies the Jewish people's right to self-determination, when it demonizes Jews, or when it holds Israel to a
double standard. This "3D test" was originally developed by Natan Sharansky in an
article for the Jerusalem Center for Public Affairs in 2004. To learn more, see "Antisemitism Uncovered Toolkit."

141  *"If one day in these United States":* "Never Is Now: Opening Remarks by ADL CEO
Jonathan Greenblatt," Anti-Defamation League, November 17, 2016, https://www.
adl.org/blog/neverisnow-opening-remarks-by-adl-ceo-jonathan-greenblatt.

143  *Questions to Ask:* Adapted from "Huddled Mass or Second Class? Challenging Anti-Immigrant Bias in the U.S.," ADL Curriculum Connections, updated May 2017, https://
www.adl.org/sites/default/files/documents/cc-huddled-mass-or-second-class.pdf, 30.
*a few ways to contribute:* Material taken from "Antisemitism Uncovered Toolkit."
*create hashtags:* Language taken from the ADL's "Confronting Hate Online," Anti-Defamation League, 2014, https://www.adl.org/sites/default/files/documents/assets/pdf/
combating-hate/confront-hate-speech-online.pdf.

144  *"unconscious attitudes, stereotypes and unintentional actions":* "Race, Perception, and Implicit
Bias," Anti-Defamation League, 2021, https://www.adl.org/education/resources/
tools-and-strategies/table-talk/race-perception-and-implicit-bias.

145  *"participants judged the black men":* "People See Black Men as Larger, More Threatening
Than Same-Sized White Men," press release, March 13, 2017, https://www.apa.org/
news/press/releases/2017/03/black-men-threatening; John Paul Wilson et al., "Racial
Bias in Judgments of Physical Size and Formidability: From Size to Threat," *Journal of
Personality and Social Psychology* 113, no. 1 (2017): 59–80.
*What language do you use:* This paragraph is adapted from the ADL's "Personal Assessment of Anti-Bias Behavior," 2007, https://www.adl.org/sites/default/files/docu
ments/assets/pdf/education-outreach/Personal-Self-Assessment-of-Anti-Bias-Behav
ior.pdf.

146  *National Socialist Movement:* For information about this group and its activities, see "National Socialist Movement," Anti-Defamation League, 2021, https://www.adl.org/
resources/backgrounders/national-socialist-movement.

147  *"Newnan believes in love for all":* "Protest Neo-Nazi Rally in Newnan, GA," GoFundMe,

last updated June 27, 2018, https://www.gofundme.com/f/protest-neonazi-rally-in-newnan-ga?utm_campaign=p_cp_url&utm_medium=os&utm_source=customer. A video of the banner is available at https://www.youtube.com/watch?v=dMvQCY A1Nkg.

*"That's not a world-changing amount of money"*: Rebecca Leftwich, "Museum Benefits from New-Nazi Rally," *Newnan Times-Herald,* June 27, 2018, https://times-herald.com/news/2018/06/african-american-museum-benefits-from-neo-nazi-rally.

148 *#Newnanstrong:* Christopher Buchanan, "#Newnanstrong Event Hopes to Counter Neo-Nazi Rally and Its Ideologies," *Alive,* April 10, 2018, https://www.11alive.com/article/news/local/newnan/newnanstrong-event-hopes-to-counter-neo-nazi-rally-and-its-ideologies/85-536661788.

*"We want there to be a different narrative"*: Buchanan, "#Newnanstrong Event."

*"It will be hard for the hate group"*: Kendall Trammell, "A Small City in Georgia Is Fighting Back Against Neo-Nazis in a Novel Way," CNN, April 21, 2018, https://www.cnn.com/2018/04/20/us/newnan-neo-nazis-rally-trnd.

*"Love Thy Neighbor"*: Jacey Fortin, "Neo-Nazi Rally Draws About Two Dozen People and Upends a Small Georgia City," *New York Times,* April 21, 2018, https://www.nytimes.com/2018/04/21/us/neo-nazi-rally-georgia.html; Najja Parker, "Hundreds of Neo-Nazis, Anti-Fascist Demonstrators and Police Were in Attendance," *Atlanta Journal-Constitution,* April 22, 2018, https://www.ajc.com/news/world/social-media-reacts-neo-nazi-rally-newnan/fsOdjsooJKiPkvnIFUys5M/.

*"Interfaith Community Unity Service"*: Shelia Poole, "Interfaith Unity Service Planned in Newnan in Response to Neo-Nazi Rally," *Atlanta Journal-Constitution,* April 20, 2018, https://www.ajc.com/lifestyles/interfaith-unity-service-planned-newnan-response-nazi-rally/9TGlWjgWyblAWsw29msukK/. Other events took place before and after the rally; see Winston Skinner, "Community Responds to Neo-Nazi Rally with Unity Gatherings," *Newnan Times-Herald,* April 19, 2018, https://times-herald.com/news/2018/04/community-responds-to-neo-nazi-rally-with-unity-gatherings.

*posted messages on social media:* Parker, "Hundreds of Neo-Nazis."

*"Those who want to fight fascism"*: "We Are #Newnanstrong," *Newnan Times-Herald,* April 22, 2018, https://times-herald.com/news/2018/04/we-are-newnanstrong.

*We are not powerless:* "We Are Not Powerless When Faced with Hate, Bias, Propaganda and Extremism," Anti-Defamation League, June 14, 2016, https://www.adl.org/blog/we-are-not-powerless-when-faced-with-hate-bias-propaganda-and-extremism.

## 8. MOBILIZING GOVERNMENT AGAINST HATE

150 *posted public warnings about it:* "'Unite the Right' Rally Could Be Largest White Supremacist Gathering in a Decade," Anti-Defamation League, August 7, 2017, https://www.adl.org/blog/unite-the-right-rally-could-be-largest-white-supremacist-gathering-in-a-decade.

*"We are stepping off"*: "Donald Trump's Failure of Character Emboldens America's Far Right," *Economist,* August 19, 2017, https://www.economist.com/united-states/2017/08/19/donald-trumps-failure-of-character-emboldens-americas-far-right.

*"very fine people on both sides"*: Jordyn Phelps, "Trump Defends 2017 'Very Fine Peo-

ple' Comments, Calls Robert E. Lee 'a Great General,'" ABC News, April 26, 2019, https://abcnews.go.com/Politics/trump-defends-2017-fine-people-comments-calls-robert/story?id=62653478.

151  "*America's white supremacist movement*": Mykal McEldowney, "What Charlottes-ville Changed," *Politico,* August 12, 2018, https://www.politico.com/magazine/story/2018/08/12/charlottesville-anniversary-supremacists-protests-dc-virginia-219353.

*galvanized public opposition:* Tom Perriello, "Very Fine People," *Slate,* August 10, 2018, https://slate.com/news-and-politics/2018/08/unite-the-right-its-legacy-will-be-the-unity-forged-by-those-on-the-other-side.html.

"*We have seen a huge number of people*": McEldowney, "What Charlottesville Changed."

*announced he would run:* "Ex–Vice President Biden 2020 Presidential Campaign Announcement," YouTube video, April 25, 2019, https://www.youtube.com/watch?v=imeLu_x0S0Y. See an account of his decision to run in Edward-Isaac Do-vere, "The Inside Story of Joe Biden's Most Fateful Decision," *Atlantic,* May 18, 2021, https://www.theatlantic.com/politics/archive/2021/05/biden-decided-run-presi dent-excerpt/618877/.

"*uphold America's values*": Joe Biden, "'We Are Living Through a Battle for the Soul of This Nation,'" *Atlantic,* August 27, 2017, https://www.theatlantic.com/politics/archive/2017/08/joe-biden-after-charlottesville/538128/.

*had not planned to run:* Graham Moomaw, "At Richmond Fundraiser, Biden Says He Wouldn't Be Running If It Weren't for Charlottesville," *Richmond Times-Dispatch,* August 27, 2019, https://richmond.com/news/local/government-politics/at-richmond-fund raiser-biden-says-he-wouldnt-be-running-if-it-werent-for-charlottesville/article_ 9558d0a9-783c-5876-9a44-9b2b7fe4653a.html.

152  *influenced by Islamist extremism:* Kristina Sgueglia, "Chattanooga Shootings 'Inspired' by Terrorists, FBI Chief Says," CNN, December 16, 2015, https://www.cnn.com/2015/12/16/us/chattanooga-shooting-terrorist-inspiration; Michael Schmidt and Jodi Rudoren, "Chattanooga Gunman Researched Islamic Martyrdom, Officials Say," *New York Times,* July 21, 2015, https://www.nytimes.com/2015/07/22/us/chattanooga-gunman-mohammod-abdulazeez.html.

"*the first global network*": https://twitter.com/Strong_Cities; "What Is the Strong Cities Network?," accessed June 18, 2021, https://strongcitiesnetwork.org/en/about/.

"*I felt the need to do something*": Andy Berke, interview with the author's research team, April 21, 2021.

*Mayor's Council Against Hate:* See the council's mission as described in the Council Against Hate, Steering Committee Report (April 11, 2019): 4, https://connect.chatta nooga.gov/wp-content/uploads/2019/04/COUNCIL-AGAINST-HATE-STEER ING-COMMITTEE-REPORT-190402-2.pdf.

154  "*good will among citizens*": Mayor William R. Wild, "Compassionate City Initiative," United States Conference of Mayors, February 2019, http://www.usmayors.org/wp-content/uploads/2019/02/WESTLAND-MI.pdf.

*Each of these initiatives:* "Best Practices," United States Conference of Mayors, 2020, https://www.usmayors.org/programs/mayors-and-business-leaders-center-for-com passionate-and-equitable-cities/best-practices/.

155  *our PROTECT Plan:* This section draws from the ADL document "PROTECT Plan to Fight Domestic Terrorism," Anti-Defamation League, accessed May 24, 2021, https://www.adl.org/protectplan.

156  *National Strategy:* "Fact Sheet: National Strategy for Countering Domestic Terror-
     ism," White House, June 15, 2021, https://www.whitehouse.gov/briefing-room/state
     ments-releases/2021/06/15/fact-sheet-national-strategy-for-countering-domestic-
     terrorism/.
     *State of Maryland:* See https://mgaleg.maryland.gov/2021RS/bills/hb/hb1227F.pdf.
158  *"war for power":* McKay Coppins, "The Man Who Broke Politics," *Atlantic,* October
     17, 2018, https://www.theatlantic.com/magazine/archive/2018/11/newt-gingrich-
     says-youre-welcome/570832/.
     *admonished political leaders:* We've also advocated against the appointment of judges and
     other officials with clear white-supremacist connections. In November 2017, for in-
     stance, we publicly opposed the confirmation of Brett Talley, who appeared to have
     spread vile Islamophobic messages and to have supported a measure to honor a founder
     of the Ku Klux Klan in online message board posts. Faced with opposition on other
     grounds, such as his woeful lack of courtroom experience, Talley subsequently and
     thankfully withdrew his nomination. See Jonathan Greenblatt, "Letter to the United
     States Senate Opposing Confirmation of Brett Talley for District Judge Post," An-
     ti-Defamation League, November 28, 2017, https://www.adl.org/news/letters/letter-
     to-the-united-states-senate-opposing-confirmation-of-brett-talley-for-district; Josh
     Gerstein and Seung Min Kim, "Two Trump Judge Nominees Out After Criticism,"
     *Politico,* December 13, 2017, https://www.politico.com/story/2017/12/13/brett-
     talley-trump-nominee-withdraws-295322.
     *forced to intervene:* Barbara Sprunt, "House Removes Rep. Marjorie Taylor Greene
     from Her Committee Assignments," NPR, February 4, 2021, https://www.npr.
     org/2021/02/04/963785609/house-to-vote-on-stripping-rep-marjorie-taylor-
     greene-from-2-key-committees.
     *Rep. Ilhan Omar:* "ADL Calls on House Leadership to Take Action After Rep. Omar's
     Anti-Semitic Tweets," Anti-Defamation League, February 11, 2019, https://www.
     adl.org/news/press-releases/adl-calls-on-house-leadership-to-take-action-after-rep-
     omars-anti-semitic.
160  *disinformation:* These definitions taken more or less verbatim from "The Dangers of
     Disinformation," Anti-Defamation League, accessed May 24, 2021, https://www.adl.
     org/education/resources/tools-and-strategies/the-dangers-of-disinformation.
161  *legally define cybercrimes:* "Virginia Man with Links to Neo-Nazis Arrested in Interna-
     tional Swatting Case," Anti-Defamation League, January 13, 2020, https://www.adl.
     org/blog/virginia-man-with-links-to-neo-nazis-arrested-in-international-swatting-
     case.
162  *a global approach:* Steve Stransky, "The 2021 NDAA, White Supremacy and Domes-
     tic Extremism," *Lawfare,* January 15, 2021, https://www.lawfareblog.com/2021-
     ndaa-white-supremacy-and-domestic-extremism.
163  *We cannot say it enough:* This chapter incorporates language and concepts in places with-
     out attribution from my January 15, 2020, testimony before the House Homeland Se-
     curity Committee, Subcommittee on Intelligence and Counterterrorism.
     *"These acts of violence":* George W. Bush, "'Islam Is Peace' Says President," White House
     Archives, September 17, 2001, https://georgewbush-whitehouse.archives.gov/news/
     releases/2001/09/20010917-11.html.
     *surged after 9/11:* Elly Belle, "Yes, 9/11 Did Cause an Increase In Islamophobia," *Refin-
     ery 29,* September 11, 2020, https://www.refinery29.com/en-us/2020/09/10019797/
     islamophobia-after-911-september-11-hate-crimes. See also Farah Pandith, *How We*

*Win: How Cutting-Edge Entrepreneurs, Political Visionaries, Enlightened Business Leaders, and Social Media Mavens Can Defeat the Extremist Threat* (New York: Custom House, 2019), 96–97.

*Sikhs:* Moni Basu, "15 Years After 9/11, Sikhs Still Victims of Anti-Muslim Hate Crimes," CNN, September 15, 2016, https://www.cnn.com/2016/09/15/us/sikh-hate-crime-victims/index.html.

164   *"When the looting starts":* Maggie Astor, "What Trump, Biden and Obama Said About the Death of George Floyd," *New York Times,* July 14, 2020, https://www.nytimes.com/2020/05/29/us/politics/george-floyd-trump-biden-obama.html.

*holding a rally:* "Trump's Appeals to White Anxiety Are Not 'Dog Whistles' — They're Racism," *Conversation,* September 18, 2020, https://theconversation.com/trumps-appeals-to-white-anxiety-are-not-dog-whistles-theyre-racism-146070.

*"President Trump's Twitter account":* "White Supremacists, Extremists Celebrate President Trump's Latest Racist Tweets," Anti-Defamation League, July 15, 2019, https://www.adl.org/blog/white-supremacists-extremists-celebrate-president-trumps-latest-racist-tweets.

165   *"we expect our politicians":* "ADL Calls on House Leadership to Take Action."

*"bombing Gaza into oblivion":* Betty McCollum, "My Bill Would Stop US Aid from Funding Palestinian Suffering," *Nation,* June 4, 2021, https://www.thenation.com/article/world/hr2590-us-aid-israel-palestine/.

166   *anti-Israel activists:* "ADL: Rutgers University Response to Anti-Semitism Has Been Insufficient," Anti-Defamation League, December 6, 2011, https://www.adl.org/news/press-releases/adl-rutgers-university-response-to-anti-semitism-has-been-insufficient. University administrations regularly fail to speak out when campus groups openly call for Zionists to be excluded from progressive spaces or from university grounds altogether; for more on this, see "Antisemitism and the Radical Anti-Israel Movement on U.S. Campuses, 2019," Anti-Defamation League, 2020, https://www.adl.org/resources/reports/antisemitism-and-the-radical-anti-israel-movement-on-us-campuses-2019.

*activist in New York City:* See "Palestine Is Still the Question: Biden, Israel, and the Fight for Palestinian Liberation," May 24, 2021, https://www.youtube.com/watch?v=yK1o1WV5EuQ. See also Sarah Ben-Nun, "Antisemite of the Year: A Law Student from New York," *Jerusalem Post,* December 23, 2020, https://www.jpost.com/diaspora/antisemitism/antisemite-of-the-year-a-law-student-from-new-york-652995.

*have often fallen silent:* "ADL: Silence from Quebec Leaders on Candidate's Anti-Semitism 'Speaks Volumes,'" Anti-Defamation League, March 31, 2014, https://www.adl.org/news/press-releases/adl-silence-from-quebec-leaders-on-candidates-anti-semitism-speaks-volumes.

167   *riven by outright extremism:* Thomas Edsall, "America, We Have a Problem," *New York Times,* December 16, 2020, https://www.nytimes.com/2020/12/16/opinion/trump-political-sectarianism.html?action=click&module=Opinion&pgtype=Homepage.

168   *"vintage fantasy":* Dorothy Wickenden, "Alexandria Ocasio-Cortez and Elizabeth Warren on the Limits of Bipartisanship," *New Yorker,* October 19, 2020, https://www.newyorker.com/podcast/political-scene/alexandria-ocasio-cortez-and-elizabeth-warren-on-the-limits-of-bipartisanship.

## 9. RAISING HATE-FREE KIDS

173  *for the Bernstein family:* For privacy reasons, I've changed names in this story and also altered certain identifying details. I draw this story from interviews my research team performed with "Gary" and "Rachel" in early May 2021.

175  *more than 80 percent affirmed:* Courtney Norris, "How Teachers Are Trying to Stop the Spread of Hate," *PBS NewsHour,* March 15, 2019, https://www.pbs.org/newshour/education/how-teachers-are-trying-to-stop-the-spread-of-hate.
   *twenty-seven hundred educators:* "Hate at School," Southern Poverty Law Center (2019): 5, https://www.splcenter.org/sites/default/files/tt_2019_hate_at_school_report_final_0.pdf.
   *remained disturbingly high:* Grant Hilary Brenner, "US High School Bullying Rates Aren't Going Down," *Psychology Today,* October 25, 2020, https://www.psychologyto day.com/us/blog/experimentations/202010/us-high-school-bullying-rates-arent-go ing-down.

176  learned *attitudes and assumptions:* See, for instance, Lisa Selin Davis, "Children Aren't Born Racist. Here's How Parents Can Stop Them from Becoming Racist," CNN, June 6, 2020, https://www.cnn.com/2020/06/06/health/kids-raised-with-bias-wellness/index.html.
   *as young as three years old:* James Burnett III, "Racism Learned," *Boston Globe,* June 10, 2012, https://www.bostonglobe.com/business/2012/06/09/harvard-researcher-says-children-learn-racism-quickly/gWuN1ZG3M40WihER2kAfdK/story.html.

177  *"parents are scared":* Robert Trestan, interview with the author's research team, April 7, 2021.

178  *conferring important intellectual and academic benefits:* Anne Fishel, "The Most Important Thing You Can Do with Your Kids? Eat Dinner with Them," *Washington Post,* January 12, 2015, https://www.washingtonpost.com/posteverything/wp/2015/01/12/the-most-important-thing-you-can-do-with-your-kids-eat-dinner-with-them/.
   *array of current topics:* I draw ideas and language in this paragraph and the next from "Table Talk: Family Conversations about Current Events," Anti-Defamation League, accessed May 24, 2021, https://www.adl.org/education/resources/tools-and-strategies/table-talk.
   *Black Lives Matter:* I draw ideas and language in this paragraph from "The Purpose and Power of Protest," Anti-Defamation League, accessed May 24, 2021, https://www.adl.org/education/resources/tools-and-strategies/the-purpose-and-power-of-protest.

179  *using music or arts projects:* See the ADL document "Talking to Young Children about Bias and Prejudice," Anti-Defamation League, accessed May 24, 2021, https://www.adl.org/education/resources/tools-and-strategies/talking-to-young-children-about-prejudice.
   *tips to keep in mind:* I've adapted these bullet points from text in the ADL pamphlet "Discussing Hate and Violence with Children," Anti-Defamation League, accessed May 24, 2021, https://www.adl.org/media/2196/download.

180  *portrayals of identity:* I draw language and ideas in this paragraph from the ADL blog's post "Diverse and Complex Narratives Cultivate Empathy and Action," Anti-Defamation League, November 10, 2020, https://www.adl.org/blog/diverse-and-complex-narratives-cultivate-empathy-and-action.

181  *lack of civics education:* Sarah Shapiro and Catherine Brown, "A Look at Civics Education in the United States," *American Educator* (Summer 2018), https://www.aft.org/ae/summer 2018/shapiro_brown.

*only about a quarter of Americans:* "Americans' Knowledge of the Branches of Government Is Declining," Annenberg Public Policy Center, September 13, 2016, https://www. annenbergpublicpolicycenter.org/americans-knowledge-of-the-branches-of-govern ment-is-declining/. Not surprisingly, trust in government is also extremely low; see "Public Trust in Government: 1958–2021," Pew Research Center, May 17, 2021, https://www.pewresearch.org/politics/2019/04/11/public-trust-in-government-1958-2019/.

182  *critical race theory:* David Theo Goldberg, "The War on Critical Race Theory," *Boston Review,* May 7, 2021, http://bostonreview.net/race-politics/david-theo-goldberg-war-critical-race-theory; Bret Stephens, "California's Ethnic Studies Follies," *New York Times,* March 9, 2021, https://www.nytimes.com/2021/03/09/opinion/californias -ethnic-studies.html?searchResultPosition=1.

183  *"as fictions constructed":* Pamela Paresky, "Critical Race Theory and the 'Hyper-White' Jew," *Sapir* 1 (Spring 2021), https://sapirjournal.org/social-justice/2021/05/critical -race-theory-and-the-hyper-white-jew/.

*doctrinaire pedagogy:* John Fensterwald, "A Final Vote, After Many Rewrites, for California's Controversial Ethnic Studies Curriculum," EdSource.com, March 17, 2021, https://edsource.org/2021/a-final-vote-after-many-rewrites-for-californias -controversial-ethnic-studies-curriculum/651338.

*"narratives of greed":* Paresky, "Critical Race Theory."

184  *"anti-bias learning environment":* Text in this and subsequent paragraphs taken from the ADL document "Creating an Anti-Bias Learning Environment," Anti-Defamation League, accessed May 21, 2021, https://www.adl.org/education/resources/tools -and-strategies/creating-an-anti-bias-learning-environment#.VGdcr_nF-So. See also Linda Santora, "How Can You Create a Learning Environment That Respects Diversity?," Anti-Defamation League, 2012, https://www.adl.org/sites/default/ files/documents/assets/pdf/education-outreach/How-Can-You-Create-a-Learning -Environment-That-Respects-Diversity.pdf.

*consider the following questions:* These questions taken from "Assessing Yourself and Your School Checklist," Anti-Defamation League, 2005, https://www.adl.org/media/ 2203/download.

188  *can happen under the radar:* This paragraph and the next one adapt text from "What Do Safe, Respectful and Inclusive Virtual Classrooms Look Like?," Anti-Defamation League, accessed May 24, 2021, https://www.adl.org/education/resources/tools-and -strategies/what-do-safe-respectful-and-inclusive-virtual-classrooms.

189  *anti-hate and antibias trainings:* See, for instance, "A Classroom of Difference," Anti-Defamation League, accessed May 24, 2021, https://www.adl.org/who-we-are/ our-organization/signature-programs/a-world-of-difference-institute/classroom; "Anti-Bias Building Blocks," Anti-Defamation League, accessed May 24, 2021, https://www.adl.org/education/resources/tools-and-strategies/anti-bias-building -blocks; "A World of Difference Institute and No Place for Hate," Anti-Defamation League, accessed May 24, 2021, https://www.adl.org/who-we-are/our-organization/ signature-programs/a-world-of-difference-institute-no-place-for-hate.

*Helping Students Make Sense of Bias:* This text box adapts text found in "Helping Students Make Sense of News Stories about Bias and Injustice," Anti-Defamation League, ac-

cessed May 24, 2021, https://www.adl.org/education/resources/tools-and-strategies/
helping-students-make-sense-of-news-stories-about-bias-and.

## 10. FAITH AGAINST HATE

194   *"respond to evil with good":* I base my account of this campaign on information contained
on the campaign's web page, "Muslims Unite for Pittsburgh Synagogue," Launch
Good, last updated January 22, 2019, https://www.launchgood.com/project/muslims
_unite_for_pittsburgh_synagogue#!/.
*support for the Jewish community:* "Emgage Condemns the Violence at the Tree of Life
Synagogue in Pittsburgh," Emgage, October 27, 2018, https://emgageusa.org/
press-release/emgage-condemns-the-violence-at-the-tree-of-life-synagogue-in
-pittsburgh/; "Pope Francis on Pittsburgh Synagogue Shooting: 'All of Us Are
Wounded by This Inhuman Act of Violence,'" CBS Pittsburgh, October 28, 2018,
https://pittsburgh.cbslocal.com/2018/10/28/pope-francis-pittsburgh-synagogue
-shooting/.

195   *Interfaith rallies and vigils:* "Interfaith Vigils Across the US for Victims of Pittsburgh
Synagogue Massacre," *VOA News,* October 28, 2018, https://www.voanews.com/
usa/interfaith-vigils-across-us-victims-pittsburgh-synagogue-massacre.
*the ADL helping to organize:* "Coming Together After the Tree of Life Synagogue Shooting," Anti-Defamation League, accessed May 24, 2021, https://www.adl.org/coming
-together-after-the-tree-of-life-synagogue-shooting.
*thousands of local citizens:* Jennifer Hijazi, "After Pittsburgh, the Interfaith Response Sends
Message of Solidarity Across the Religious Divide," *Vox,* October 28, 2018, https://
www.vox.com/2018/10/28/18034460/pittsburgh-shooting-interfaith-response-vigil
-fundraising.
*Buddhist monks visited the synagogue:* Allie Yang, "'We Will Be There for Them in Any
Way We Can': Pittsburgh Islamic Center Head on Standing with Jewish Community
in Wake of Synagogue Shooting," ABC News, October 29, 2018, https://abcnews.go
.com/US/pittsburgh-islamic-center-head-standing-jewish-community-wake/story?id
=58837581.
*group of African immigrants:* Beth Kissileff and Eric Lidji, eds., *Bound in the Bond of Life:
Pittsburgh Writers Reflect on the Tree of Life Tragedy* (Pittsburgh: University of Pittsburgh
Press, 2020), 128–29.
*"that's the most powerful story":* Rabbi Hazzan Jeffrey Myers, interview with the author's
research team, March 15, 2021.

196   *gunned down:* Yossi Klein Halevi, "The War on the Israeli Home Front," *Wall Street
Journal,* May 18, 2014, https://www.wsj.com/articles/yossi-klein-halevi-the-war-on
-the-israeli-home-front-1416355411; Isaac Keiser, "Israeli Democracy Still Haunted
by the Ghosts of Meir Kahane and Baruch Goldstein," Foreign Policy Research Institute, September 3, 2019, https://www.fpri.org/article/2019/09/israeli-democracy
-still-haunted-by-the-ghosts-of-meir-kahane-and-baruch-goldstein/.
*the ADL called out:* "ADL: Israeli Chief Rabbi Statement Against Non-Jews Living
in Israel Is Shocking and Unacceptable," Anti-Defamation League, March 28, 2016,
https://www.adl.org/news/press-releases/adl-israeli-chief-rabbi-statement-against
-non-jews-living-in-israel-is-shocking.

197   *"If we don't speak up":* "Interfaith Dialogue: A Jew and a Muslim Discuss the Spate of

Hate Crimes in Aftermath of 2016 Election," Anti-Defamation League, December 21, 2016, https://www.adl.org/blog/interfaith-dialogue-a-jew-and-a-muslim-discuss-the -spate-of-hate-crimes-in-aftermath-of-2016.

*has plummeted:* Jeffrey M. Jones, "U.S. Church Membership Falls Below Majority for First Time," Gallup, March 29, 2021, https://news.gallup.com/poll/341963/church -membership-falls-below-majority-first-time.aspx.

198   *long pursued interfaith relationships:* I draw language in this paragraph from "Interfaith Coalition on Mosques, Statement of Purpose," Anti-Defamation League, 2015, https://www.adl.org/media/6940/download; "Interfaith Coalition on Mosques," Anti-Defamation League, accessed May 24, 2021, https://www.adl.org/interfaith -coalition-on-mosques; and "Interfaith and Intergroup Relations," Anti-Defamation League, accessed May 24, 2021, https://www.adl.org/what-we-do/promote-respect/ interfaith.

*We also hold events:* For example, "ADL Event on Interreligious Tolerance and the UAE," Anti-Defamation League, June 7, 2019, https://www.adl.org/blog/adl-event -on-interreligious-tolerance-and-the-uae; "Next Year May We ALL Be Free," An-ti-Defamation League, April 7, 2017, https://www.adl.org/news/article/next-year -may-we-all-be-free.

*"The support and understanding":* This section is an adaptation of Lorraine Array, "The Ripple Effect of Interfaith Dialogue," Anti-Defamation League, March 23, 2015, https://www.adl.org/blog/the-ripple-effect-of-interfaith-dialogue. Portions of the language used in this section first appeared there.

200   *"What I do":* Abdullah Antepli, interview with the author's research team, May 18, 2021.

*his own story:* I draw this account of Antepli's childhood from my informal conversations with him, from an interview with him conducted for this book on May 18, 2021, and from a recorded interview of him speaking during a class at Duke University. All quotes from Antepli over the next several paragraphs come from these sources.

203   *"I felt my legs were praying":* Susannah Heschel, "Their Feet Were Praying," *Times of Israel,* January 10, 2012, https://jewishweek.timesofisrael.com/their-feet-were -praying/.

*"Praying for freedom":* Rabbi Laurie Green, "Praying with Our Feet," T'ruah, 2019, https://truah.org/resources/praying-with-our-feet/.

205   *anti-hate legislation:* Dennis Romboy, "Church Will Not Oppose Hate Crimes Legislation, Clarifies Previous Position," *Deseret News,* January 23, 2019, https://www .deseret.com/2019/1/23/20664024/church-will-not-oppose-hate-crimes-legislation -clarifies-previous-position.

206   *a 2012 massacre:* Steven Yaccino, Michel Schwirtz, and Marc Santora, "Gunman Kills 6 at a Sikh Temple Near Milwaukee," *New York Times,* August 5, 2012, https://www .nytimes.com/2012/08/06/us/shooting-reported-at-temple-in-wisconsin.html.

*"We feel compelled":* "Tree of Life Stands with the Christchurch Mosques," GoFundMe, last updated June 19, 2019, https://www.gofundme.com/f/tree-of-life-stands-with -christchurch-mosques?utm_campaign=p_cp_url&utm_medium=os&utm_source= customer.

*Members of the congregation:* Alex Horton, "Their Fellow Congregants Died in Pittsburgh. Now Jews Are Supporting Muslims in New Zealand," *Washington Post,* March 18, 2019, https://www.washingtonpost.com/religion/2019/03/18/their-congregants -died-pittsburgh-now-jews-are-supporting-muslims-new-zealand/.

## 11. BUILDING BETTER BUSINESSES

210   *long lines at polling places:* Sam Levine and Agencies, "More Than 10-Hour Wait and
      Long Lines as Early Voting Starts in Georgia," *Guardian,* October 12, 2020, https://
      www.theguardian.com/us-news/2020/oct/13/more-than-10-hour-wait-and-long
      -lines-as-early-voting-starts-in-georgia.
      *credited with increasing turnout:* Trone Dowd, "Georgia Republicans Just Made It a Crime
      to Give Voters Snacks," *Vice,* March 25, 2021, https://www.vice.com/en/article/
      88az7k/georgia-republicans-want-to-make-it-a-crime-to-give-voters-snacks-while
      -they-wait.
      *one of hundreds:* Amy Gardner, Kate Rabinowitz, and Harry Stevens, "How GOP-
      Backed Voting Measures Could Create Hurdles for Tens of Millions of Voters," *Wash-
      ington Post,* March 11, 2021, https://www.washingtonpost.com/politics/interactive/
      2021/voting-restrictions-republicans-states/.
      *"the Christmas tree of goodies":* Kelly Mena et al., "Georgia Republicans Speed Sweeping
      Elections Bill Restricting Voting Access into Law," CNN, March 26, 2021, https://
      www.cnn.com/2021/03/25/politics/georgia-state-house-voting-bill-passage/index
      .html.
      *"the most pernicious thing":* Nick Corasaniti, "Georgia G.O.P. Passes Major Law to Limit
      Voting Amid Nationwide Push," *New York Times,* March 25, 2021, https://www
      .nytimes.com/2021/03/25/us/politics/georgia-voting-law-republicans.html.
      *"undoubtedly disenfranchise thousands":* "ADL Opposes Georgia Voting Suppression
      Bills," Anti-Defamation League, February 26, 2021, https://atlanta.adl.org/news/adl
      -opposes-georgia-voting-suppression-bills/.
211   *shameful efforts made by racist legislatures:* The passage of the Fifteenth Amendment in 1870
      forbade states from infringing on the rights of Black citizens to vote, but during the
      decades afterward many states passed laws that in one way or another did precisely
      that. Some of these laws said you couldn't vote unless your grandfather had, thus dis-
      qualifying descendants of freed slaves. Some said citizens had to pay a special tax in
      order to vote—difficult or impossible for many poor Blacks. Some required citizens
      to pass absurd and subjective literary tests that administrators applied to keep illiterate
      Blacks out of the electorate but not whites. See, for instance, "Voting Rights for Afri-
      can Americans," Library of Congress, accessed May 18, 2021, https://www.loc.gov/
      classroom-materials/elections/right-to-vote/voting-rights-for-african-americans/;
      "Poll Taxes," National Museum of American History, Behring Center, accessed May
      18, 2021, https://americanhistory.si.edu/democracy-exhibition/vote-voice/keeping
      -vote/state-rules-federal-rules/poll-taxes; "Civil Rights in America: Racial Voting
      Rights," National Park Service, U.S. Department of the Interior, National Historic
      Landmarks Program (2009): 12–13, https://www.nps.gov/subjects/tellingallamerica
      nsstories/upload/CivilRights_VotingRights.pdf.
      *famed march from Selma to Montgomery:* "A Look Back: ADL's Role in Selma and the Vot-
      ing Rights Act," Anti-Defamation League, February 24, 2015, https://www.adl.org/
      news/article/a-look-back-adls-role-in-selma-and-the-voting-rights-act.
      *put an end to:* "A History of Voter Suppression," National Low Income Housing Co-
      alition, September 23, 2020, https://nlihc.org/resource/history-voter-suppression;
      https://www.nationalgeographic.com/history/article/voter-suppression-haunted
      -united-states-since-founded.

*spoken out against the new voter-suppression measures:* "Safeguarding the Right to Vote," Anti-Defamation League, August 20, 2013, https://www.adl.org/news/article/safeguarding-the-right-to-vote; Matt DeRienzo, "Analysis: New and Age-Old Voter Suppression Tactics at the Heart of the 2020 Power Struggle," Center for Public Integrity, October 28, 2020, https://publicintegrity.org/politics/elections/ballotboxbarriers/analysis-voter-suppression-never-went-away-tactics-changed/. For a list of voter-suppression measures, see "61 Forms of Voter Suppression," Voting Rights Alliance, accessed May 21, 2021, https://www.votingrightsalliance.org/forms-of-voter-suppression.

*took a pass on civic activism:* Andrew Ross Sorkin et al., "More Businesses Are Standing for Justice," *New York Times,* April 21, 2021, https://www.nytimes.com/2021/04/21/business/dealbook/business-civil-rights-george-floyd.html.

*hesitation began to change:* Melissa Repko et al., "Hashtags Won't Cut It. Corporate America Faces a Higher Bar in a Reckoning on Racial Inequality," CNBC, June 12, 2020, https://www.cnbc.com/2020/06/12/action-wanted-corporate-america-faces-a-higher-bar-on-racial-inequality.html.

*stayed noticeably silent:* David Gelles, "Corporations, Vocal About Racial Justice, Go Quiet on Voting Rights," *New York Times,* April 5, 2021, https://www.nytimes.com/2021/03/29/business/corporate-america-voting-rights.html.

212  *"take a nonpartisan stand":* "Black Executives Call on Corporations to Fight Restrictive Voting Laws," Black Economic Alliance, March 31, 2021, https://blackeconomicalliance.org/news/black-executives-call-on-corporations-to-fight-restrictive-voting-laws/. See also Todd Frankel, "How Two Black CEOs Got Corporate America to Pay Attention to Voting Rights," *Washington Post,* May 4, 2021, https://www.washingtonpost.com/business/2021/05/04/corporate-america-voting-rights/.

*hundreds of companies:* Todd C. Frankel, Josh Dawsey, and Jena McGregor, "Mounting Corporate Opposition to Proposed Voting Restrictions Tests Long-Standing Alliance with GOP," *Washington Post,* April 14, 2021, https://www.washingtonpost.com/business/2021/04/14/letter-voting-georgia-buffet-target-netflix-amazon/; David Gelles and Andrew Ross Sorkin, "Hundreds of Companies Unite to Oppose Voting Limits, but Others Abstain," *New York Times,* April 14, 2021, https://www.nytimes.com/2021/04/14/business/ceos-corporate-america-voting-rights.html; Todd Frankel, "More Than 100 Corporate Executives Hold Call to Discuss Halting Donations and Investments to Fight Controversial Voting Bills," *Washington Post,* April 11, 2021, https://www.washingtonpost.com/business/2021/04/11/companies-voting-bills-states/.

*2021 All-Star game:* "MLB to Move All-Star Game Out of Atlanta Over Georgia Voting Restrictions," *Axios,* April 2, 2021, https://www.axios.com/mlb-all-star-game-georgia-voting-restrictions-acfc5b02-83d1-440a-9367-4295d01561ed.html.

*"getting in bed with the Left":* Frankel, Dawsey, and McGregor, "Mounting Corporate Opposition."

*"stay out of politics":* Matthew Brown, "McConnell to CEOs: 'Stay Out of Politics.' Republicans Threaten Businesses Opposing Georgia Voting Law," *USA Today,* April 6, 2021, https://www.usatoday.com/story/news/politics/2021/04/06/mitch-mcconnell-threatens-corporations-over-georgia-voting-law/7103439002/.

*"all of the woke companies":* Oriana Gonzalez, "Trump Calls for MLB Boycott After All-Star Game Moves Out of Georgia," *Axios,* April 3, 2021, https://www.axios.com/

trump-major-league-baseball-boycott-georgia-voting-34e052f0-89eb-4cc7-8a99
-d63bd8709077.html.

213 *boycott North Carolina:* Marisa Taylor, "Inside Corporate America's Stand Against
Transgender Discrimination," *Guardian,* October 1, 2016, www.theguardian.com/
sustainable-business/2016/oct/01/north-carolina-hb2-law-transgender-issues
-corporate-businesses-protest.

*people trust business:* "Edelman Trust Barometer 2021," Edelman, last updated March 16,
2020, https://www.edelman.com/trust/2021-trust-barometer.

*"Business has to help the country":* Kathryn Dill and Kurt Wilberding, "More Trust in
Business Than in Government and Media, Survey Finds," *Wall Street Journal,* January
13, 2021, https://www.wsj.com/articles/more-trust-in-business-than-in-government
-and-media-survey-finds-11610533801.

214 *"the social responsibility of business":* Friedman, "A Friedman Doctrine."

215 *made businesses legally accountable:* "Certification," Certified B Corporation, accessed June
20, 2021, https://bcorporation.net/.

216 *"a fundamental commitment":* "Business Roundtable Redefines the Purpose of a Corporation to Promote 'an Economy That Serves All Americans,'" Business Roundtable, August 19, 2019, https://www.businessroundtable.org/business-roundtable
-redefines-the-purpose-of-a-corporation-to-promote-an-economy-that-serves-all-
americans.

217 *have diverse workforces:* Robin J. Ely and David A. Thomas, "Getting Serious About Diversity: Enough Already with the Business Case," *Harvard Business Review* (November/
December, 2020), https://hbr.org/2020/11/getting-serious-about-diversity-enough
-already-with-the-business-case.

218 *large majorities of Americans:* Repko et al., "Hashtags Won't Cut It."

*almost 80 percent of those surveyed:* "2021 Edelman Trust Barometer: Business and Racial Justice in America," Edelman, May 2021, https://www.edelman.com/trust/2021
-trust-barometer/business-racial-justice.

*Research published in 2018:* "Social Accountability: The Business World's New Role in
Social Change," *Business News Daily,* March 3, 2020, https://www.businessnewsdaily
.com/10487-corporate-social-accountability.html.

*A 2020 study found:* "The Corporate Social Mind Research Report," Engage for
Good, June 2020, https://engageforgood.com/the-corporate-social-minds-2020
-public-expectations-of-companies-to-address-social-issues/.

*over 80 percent in one survey:* Porter Novelli, "73 Percent of Business Executives Agree
Companies Have More Responsibility Than Ever Before to Take Stands on Social Justice Issues, According to Research by Porter Novella [*sic*]," *Cision,* September 17, 2020,
https://www.prnewswire.com/news-releases/73-percent-of-business-executives
-agree-companies-have-more-responsibility-than-ever-before-to-take-stands-on
-social-justice-issues-according-to-research-by-porter-novella-301133383.html.

219 *can hold these services accountable:* Some language in this paragraph and elsewhere in this
chapter originally appeared in a speech I gave to the World Federation of Advertisers on April 20, 2021. For the full text of the speech, see "Jonathan Greenblatt's Remarks to the World Federation of Advertisers," Anti-Defamation League, April 20,
2021, https://www.adl.org/news/article/jonathan-greenblatts-remarks-to-the-world
-federation-of-advertisers.

221 *he even targeted me:* Aaron Bandler, "Farrakhan Calls ADL CEO 'Satan' in July 4 Ad-

dress," *Jewish Journal,* July 6, 2020, https://jewishjournal.com/news/united-states/318512/farrakhan-calls-adl-ceo-satan-in-july-4-address/.

224    *anti-Israel sentiments:* Alana Goodman, "Meet Ben & Jerry's Board Chair: Anti-Israel Activist Has Published Defenses of Hezbollah, Hamas," *Washington Free Beacon,* July 23, 2021, https://freebeacon.com/national-security/meet-ben-jerrys-board-chair-anti-israel-activist-has-published-defenses-of-hezbollah-hamas/.

    *explicitly informed me:* Lazar Berman, "After Ben & Jerry's Controversy, Unilever Tells ADL Chief: We Don't Support BDS," *Times of Israel,* July 27, 2021, https://www.timesofisrael.com/liveblog_entry/after-ben-jerrys-controversy-unilever-tells-adl-chief-we-dont-support-bds/.

    *a middle ground:* Bennett Cohen and Jerry Greenfield, "We're Ben and Jerry. Men of Ice Cream, Men of Principle," *New York Times,* July 28, 2021, https://www.nytimes.com/2021/07/28/opinion/ben-and-jerry-israel.html.

    *prospect of legal action:* Jacob Magid, "US States' Fully Baked Anti-BDS Laws Could Put the Freeze on Ben & Jerry's," *Times of Israel,* July 21, 2021, https://www.timesofisrael.com/us-states-fully-baked-anti-bds-laws-could-put-the-freeze-on-ben-jerrys/.

227    *pledges from companies:* Andrew Ross Sorkin and Edmund Lee, "Asian-American Business Leaders Fund Effort to Fight Discrimination," *New York Times,* May 3, 2021, https://www.nytimes.com/2021/05/03/business/dealbook/asian-american-donation-philanthropy.html; "Asian American Foundation Launched with Pledges Totaling $250 Million," *Philanthropy News Digest,* May 4, 2021, https://philanthropynewsdigest.org/news/asian-american-foundation-launched-with-pledges-totaling-250-million; "Our Partners," Asian American Foundation, accessed May 20, 2021, https://www.taaf.org/our-partners.

    *powerful mission:* "Our Mission," Asian American Foundation, https://www.taaf.org/our-mission#:~:text=To%20serve%20the%20Asian%20American,discrimination%2C%20slander%2C%20and%20violence.

## EPILOGUE

229    *"I have surely seen the affliction":* The book of Shemot (Exodus), chapter 3, Jewish Virtual Library, https://www.jewishvirtuallibrary.org/shemot-exodus-chapter-3.

231    *countless inspiring stories:* I draw language and content in this paragraph from "ADL Impact Stories, Organized by Topic," Anti-Defamation League, accessed June 18, 2021, https://www.adl.org/impact-stories.

# INDEX

Page numbers in *italics* indicate illustrations; "n" indicates a note.

Anti-Defamation League (ADL) (*cont.*)
  mission, 3–4, 140–42
  missteps, 141–42
  Never Is Now summit, 15, 100, 141
  No Place for Hate program, 189
  online harassment survey, 5
  parenting resources, 177–78
  PROTECT Plan, 155–62
  Pyramid of Hate, 43–46, *44*
  radicalization, interruption of, 12
  sharing facts, 138
  SJP attack, 35
  Stop Hate for Profit, 99
  "Table Talk" conversation guides, 177
  tracking white-supremacist propaganda,
    49
  two-state solution, 84
  voting rights, 210–11
Antifa, tactics, 148
anti-government activists, 36
anti-hate efforts. *See* hate, pushback against
anti-immigrant sentiment
  Chinese massacre, Los Angeles, 29
  "great replacement theory," 63, 129–31
  Soros rumors, 32
  Trump's, 48
  Trump years, 27
  white-nationalist ideology, 27
anti-Israel sentiment. *See also* anti-Zionism;
    Boycott, Divestment, Sanctions
  antisemitism and, 118, 166–67, 264n139
  college campuses, 166
  dehumanizing Jews, 49–50
  Far Left, 117–18, 164–65
  mainstreaming of, 118
  Obama White House, 254n86
  United States, 34–35
anti-Muslim sentiment, 27, 48, 141, 163,
    199. *See also* Christchurch, New
    Zealand
antisemitic incidents. *See also* synagogue
    attacks
  college campuses, 8–9
  Europe, 118
  hate crimes against Jews, 49
  increase in, 4, 9, 118
  introduction, 3–5
  Los Angeles attack, 9

Miami, 9
New York City attack, 9
number of, 27
online gaming, 40
in schools, 191–92
support for Palestinian cause and, 9
United Kingdom, 118
United States, 118
Zoom-bombing, 92–93
antisemitism
  anti-Israel sentiment and, 118, 166–67,
    264n139
  anti-Zionism as, 4, 8–9, 28–29, 34,
    264n139
  conspiracy theories, 97–98, 129
  Facebook posts, 96
  France, 23
  increase in, 9, 49
  Iran, 196
  Islamic rhetoric, 199–202
  left's failure to condemn, 9–10, 243n9
  Louis Farrakhan, 220–21
  Middle East, 29
  Nazis, 5
  normalization of, 6, 35–36
  from political left, 28, 35
  Russian history, 29
  schoolchildren, 173–75
  in schools, 183
  Soviet Union, 29
  Spanish Inquisition, 5
  as staple of American life, 3
  support for Palestinian cause and, 243n9
  "3D test," 264n139
  triggering violence, 5
  tropes, 32, 34, 97–98, 139, 165
  U.S. Congress, 4, 158–59, 164–65
  U.S. history, 12–13
  Voltaire, 244n15
  white-nationalist ideology, 27
anti-Zionism. *See also* anti-Israel sentiment;
    Boycott, Divestment, Sanctions
  as antisemitism, 4, 8–9, 28–29, 34
  cancel culture and, 86–88
  college campuses, 88, 166, 268n166
  countering hate, 138
  dehumanization of Jews, 49–50
  Facebook posts, 117–18

Israel (*cont.*)
  hostility toward, as antisemitism, 4–5, 264n139
  Iran's demonization of, 54–55
  normalization of hate against, 32
  Palestinian intifada (1980s), 201
  Turkey alliance, 142
  two-state solution, 84, 224–25
  U.S. embassy, 81–82
Italian-Canadians, 121
*It Can't Happen Here* (Lewis), 11
It Started with Words (Claims Conference campaign), 248n43

January 6, 2021 attack on U.S. Capitol, 4, 8, 36–37, 102, 112, 120
Japanese-Americans, 121
Japanese-Canadians, 121
Jersey City, New Jersey, 4, 25
Jerusalem, 81–82, 202
Jewish extremists, 196
Jewish Museum, Brussels, Belgium, 19
"the Jewish Question," 5
Jewish Voice for Peace (JVP), 34
Jews. *See also* antisemitism; Holocaust; Israel
  civil rights movement, 196, 203
  in conspiracy theories, 97–98, 129
  faith against hate, 196–97
  Far Left's dehumanization of, 49–50
  hate crimes against, 49
  historical trauma, 116–17
  housing discrimination against, 140
  interfaith actions, 196, 198–99, 202, 206–7
  Iranian revolution (1979), 52–55
  as Nazi undesirables, 5, 121–22
  Passover holiday, 173–74
  social media posts targeting, 96
  taking nothing for granted, 116–17
  *teshuvah* (repentance), 78
  violence against, 9, 19–25, 124, 230 (*See also* synagogue attacks)
Johnson, Lyndon B., 141
Johnson and Johnson, 104
Jones, Alex, 33, 97, 99
Jope, Alan, 224

J Street (Zionist advocacy organization), 84–85
Judaism. *See* Jews
JVP (Jewish Voice for Peace), 34

Kahane, Meir, 196
Kaine, Tim, 151
Karamehic-Oates, Adna, 56–57, 58, 60
Kardashian, Kim, 101
Kelly, John, 21
Kendi, Ibram X., 182
Kennedy, John F., 141
Kennedy, Robert, 141
Khamenei, Ayatollah Ali, 103
Khomeini, Ayatollah, 53, 54, 97
kids. *See* children; schools
KIND Snacks, 215
King, Martin Luther, Jr., 14, 140–41, 196, 203, 211
King, Steve, 158
Kiswani, Nerdeen, 166
KKR (company), 226
Kosovo, 57, 58
Ku Klux Klan (KKK), 27, 65, 73, 74, 267n158

Labour Party (UK), 6
Latinx people, 25, 68–69, 119, 129
Lavi, Gorgi, 53
leadership
  corporate responsibility, 218–20, 224, 227
  faith against hate, 205–9
  government, 162–67
left-wing extremism, 36. *See also* Far Left
Leonard, Meyers, 40–43, 45, 51
letter-writing, 137, 143
Levitsky, Steven, 122
Lewis, Sinclair, 11
LGBTQ people
  countering hate against, 138
  hate-crime legislation and, 204–5
  as Nazi undesirables, 5
  systemic discrimination against, 46
  transgender bathroom laws, 213, 223
  violence against, 26, 206, 245n26
Li Lu, 226
Lincoln, Abraham, 114–15, 125